I0803180

Scalable Video Streaming with Fountain Codes

Copyright and Permission Notice

© 2011 IEEE. Portions reprinted, with permission, from S. Nazir, D. Vukobratović, and V. Stanković, "Performance evaluation of Raptor and Random Linear Codes for H.264/AVC video transmission over DVB-H networks," ICASSP-2011 IEEE Intl. Conf. Acoustics Speech and Sig. Proc., Prague, Czech Republic, May 2011, Pages: 2328 - 2331.

© 2011 IEEE. Portions reprinted, with permission, from S. Nazir, V. Stanković, and D. Vukobratović, "Unequal error protection for data partitioned H.264/AVC video streaming with Raptor and Random Linear Codes for DVB-H networks," ICME-2011 IEEE Intl. Conf. Multimedia and Expo, Barcelona, Spain, July 2011, Pages: 1 - 6.

© 2011 IEEE. Portions reprinted, with permission, from S. Nazir, V. Stankovic, and D. Vukobratovic, "Expanding Window Random Linear Codes for Data Partitioned H.264 Video Transmission over DVB-H Network," ICIP-2011 IEEE Intl. Conf. on Image Processing, Brussels, Belgium, September 2011, Pages: 2205 - 2208.

© 2011 IEEE. Portions reprinted, with permission, from S. Nazir, V. Stanković, and D. Vukobratović, "Scalable broadcasting of sliced H.264/AVC over DVB-H network," ICON-2011 IEEE Intl. Conf. on Networks, Special Session on Robust and Scalable Multimedia Networking, Singapore, December 2011, Pages: 36 - 40.

© 2012 IEEE. Portions reprinted, with permission, from S. Nazir, V. Stankovic, and D. Vukobratovic, "Adaptive layered multiple description coding for wireless video with Expanding Window Random Linear Codes," ICASSP-2012 IEEE Intl. Conf. on Acoustic, Speech, and Signal Processing, Kyoto, Japan, March 2012, Pages: 1353 - 1356.

© 2013 IEEE. Portions reprinted, with permission, from S. Nazir, V. Stankovic, H. Attar, L. Stankovic, and S. Cheng,"Relay-assisted Rateless Layered Multiple Description Video Delivery," IEEE Journal on Selected Areas in Communications, August 2013, Pages: 1629 - 1637.

Portions reprinted, with permission, from S. Nazir, D. Vukobratović, V. Stanković, and I. Andonović, "Rate Adaptive Selective Segment Assignment for Reliable Wireless Video Transmission," International Journal of Digital Multimedia Broadcasting, vol. 2012, Article ID 160521, 2012.

Portions reprinted, with permission, from S. Nazir, F. Verdicchio, and G. Fairhurst, "Scalable Video Streaming Techniques over Cooperative Relay Networks", International Journal of Computer Applications, Vol.66 2013.

Portions reprinted with kind permission from Springer Science+Business Media: "Unequal Error Protection for Data-partitioned H.264/AVC Video Broadcasting", Springer Multimedia Tools and Applications, 2014, Pages: 5787-5809, S. Nazir, D. Vukobratovic, V. Stankovic, I. Andonovic, K. Nybom, and S. Gronroos,.

Scalable Video Streaming with Fountain Codes

By

Sajid Nazir

Cambridge
Scholars
Publishing

Scalable Video Streaming with Fountain Codes

By Sajid Nazir

This book first published 2016

Cambridge Scholars Publishing

Lady Stephenson Library, Newcastle upon Tyne, NE6 2PA, UK

British Library Cataloguing in Publication Data
A catalogue record for this book is available from the British Library

Copyright © 2016 by Sajid Nazir

All rights for this book reserved. No part of this book may be reproduced, stored in a retrieval system, or transmitted, in any form or by any means, electronic, mechanical, photocopying, recording or otherwise, without the prior permission of the copyright owner.

ISBN (10): 1-4438-7757-3
ISBN (13): 978-1-4438-7757-2

I dedicate this book to my brother, Shahid Nazir, who initiated me on the journey of image and video coding and inspired me to learn computer programming.

TABLE OF CONTENTS

FOREWORD

The area of video streaming has seen tremendous growth due to the enhanced processing power, better compression algorithms and increased bandwidth in the emerging networks. However research challenges exist to overcome effects of video packet losses and delays to provide a better user experience. Fountain codes are used in this book to address some of the challenges in this research area, and are combined in innovative ways with the different importance classes of compressed video data. This book represents a useful reference point for researchers, academics, research students, and industry developers interested in utilizing the error correction codes for ensuring better video quality.

The book is organized in five chapters. The outline of the chapters is as follows:

Chapter 1 **Introduction to Video Communications and Streaming** describes the basic techniques employed for the encoding of video data and the latest video coding standards. The error resilience features present in the latest video coding standards, Multi Description Coding (MDC), transport protocols and multimedia communication standards are also described. Some applications of video streaming are also provided.

Chapter 2 **Fountain Codes for Error Correction** covers the need for the error correction codes and demonstrates their use for protecting the video data. State-of-the-art fountain coding schemes namely LT codes, Raptor codes and Random Linear Codes are described. An application of Fountain codes is presented for DVB-H network.

Chapter 3 **Rate Adaptation Techniques for Video Streaming** highlights the importance for the video application to adapt its rate to the underlying channel bandwidth. This chapter proposes a novel adaptive scheme, called RASSA, for unicast video transmission.

Chapter 4 **Video Delivery over Wireless Relay Networks** covers the cooperative relay based multi hop wireless network to symbolize the 4G communication standard, LTE-A. The transmission of layered video data for H.264/AVC and H.264/SVC is compared over relay networks.

A multiple description coding (MDC) scheme based on the slicing and data partitioning (DP) is proposed. Expanding window-random linear codes are used to simulate the transmission of the layered descriptions for LTE-A standard.

Chapter 5 **Future of Video Streaming** points to some of the promising technological developments for video streaming.

PREFACE

Most of the latest communication standards are Internet protocol (IP) based, whereas Internet provides only a best-effort service model and the priority-based service models are only gradually being realized for real-time data. Multimedia communication over such wireless channels is difficult due to the fact that the channel conditions are generally poor and also the channel characteristics can change very rapidly. Considering the importance of the issues highlighted above, this book focuses on designing techniques to exploit different importance classes in compressed video data to design adaptive solutions to support multimedia traffic over wireless channels.

The video content is sent and received in compressed form as small packets and the video playback can start after some data has been buffered at the receiver. Thus, video streaming enables playing of video as more data is still being received and the user does not have to wait for the entire file to be downloaded.

The video streaming over IP networks has seen tremendous growth and this traffic is likely to increase as new platforms and applications evolve. The area of video communications over current heterogeneous networks is very important and challenging. Internet was never designed to work with such real-time traffic. Considering a TCP/IP network, the receipt of each packet is acknowledged to ensure the guaranteed delivery but this works against video data characteristics where it can tolerate some loss but not the delay in delivery. Mobile wireless channels and IP based communications are inherently prone to errors and packet losses which highlight the need for better error protection and recovery mechanisms.

Error resilience features and Forward Error Correction (FEC) at the application layer are often used to protect the video data against losses. One class, Fountain codes are rateless codes which can be used to potentially generate an unlimited number of encoded packets from a limited number of source packets. The decoding is possible if the number of received encoded packets at the receiver is just a little more than the source packets. The delivery and recovery of the original encoded information thus is independent of the network reliability. However, there is a small penalty to be paid for the encoding and decoding complexity.

As each portion of compressed video data does not have equal importance for the video re-construction, this characteristic can also be advantageously exploited while designing FEC solutions by providing more protection to more important portions. Thus efficient FEC solutions can be designed which take advantage of the scalable nature of current video standards and the unequal importance of video data. The amount of video data to be transmitted and the degree of protection by the rateless codes can be adaptively applied to match the channel conditions and is important for transmission to heterogeneous receivers.

Scalable video coding has come into prominence with H.264 Scalable Video Coding (SVC) standard. Scalable video means that it is possible to extract a subset of compressed video data which can be used to playback the video at a lower quality. The advantage with the use of scalable video coding is that the data gets coded as an important base layer, sufficient to create a low quality representation, and one or more enhancement layers, each of which refine the video quality. This property of video data makes it possible to use the fountain codes to provide Unequal Error Protection (UEP), that is, more protection to the more important data. UEP could also be used for other importance classes that can be realised in AVC standard through data partitioning or slices or by assigning a higher protection to more important intra (I) frame as compared to P and B frames.

In order to ensure quality of experience, FEC is the scheme proposed in this book. Fountain codes have emerged as a dominant scheme where the degree of protection does not depend on which packets are received but rather how many are received. This characteristic of fountain codes can result in many simplifications to the data transport techniques.

The video coding standards H.264/AVC and SVC together with the latest standard High Efficiency Video Coding (HEVC) provide mechanisms to match the video content to the available channel bandwidth through selecting a subset of compressed video for transmission with resulting reconstruction at a reduced quality. It is thus imperative to design FEC solutions which are adaptive to the varying wireless channel conditions, i.e., bandwidth and packet loss rate. This adaptive behaviour becomes even more important for transmission to heterogeneous receivers.

The book covers the Luby Transform (LT codes), Random Linear Codes (RLC) and Raptor coding based techniques for the mobile television broadcasting standards like Digital Video Broadcasting-Handheld (DVB-H) and DVB-T2 (Second Generation Terrestrial). A reliable unicast video communication solution based on LT codes is proposed by exploiting Unequal Error Protection (UEP) for encoded video data partitioned with the Data Partitioning (DP) and slicing feature of H264/AVC. A comparison between H.264/AVC and H.264/SVC data is given for relay collaboration strategies, namely Amplify-and-Forward (AF) and Decode-and-Forward (DF). The possible applications and advantages of each scheme are highlighted. A novel solution for Multiple Description Coding (MDC) is proposed based on DP and slicing features of H.264/AVC. The proposed solution is distinct because within each description created by slicing feature; provision of exploiting UEP is also available by utilizing the DP feature. Thus many possibilities emerge to selectively transmit the partitions of each description to exploit the multi path diversity.

The book describes how the scalable and error-resilience features of the video coding standards can be optimally combined with fountain codes to ensure quality of video under given channel conditions. The proposed solutions are extensible to the other video coding standards and error correction codes. The book aims to present simple and generalized solutions which can be extended to the latest video compression standards, such as HEVC. For example, the approach can be extended to other error-resilience features or importance classes within compressed video data.

ACKNOWLEDGEMENTS

The idea to write this book arose during my PhD when I developed a passion for video communications. I am thankful to the reviewers and publishers for enabling me to present my work as a book.

I am extremely thankful for the support and guidance provided by my PhD advisor Dr. Vladimir Stankovic, during my PhD. I am highly indebted to Dr. Dejan Vukobratovic for the guidance I received from him during his tenure at University of Strathclyde. I acknowledge his help for providing me with the exposure and understanding of fountain codes, and providing the code for optimizing probabilities. I am thankful to Dr. Kristian Nybom for providing the DVB error trace files and guidance for DVB-H and DVB-T2 systems. I acknowledge the support of Dr Samuel Cheng who provided the dynamic programming solution for the optimization problem for relay networks.

I am greatly indebted to my wife for sparing me the long hours that I spent writing the book. I am also thankful to my children, Maria, Salman and Haider, who showed a lot of understanding and maturity in keeping with the long hours of my PhD, and then in the writing of this book.

About the Author

Sajid Nazir received his BE degree in Electrical Engineering from NED University, Karachi, Pakistan in 1986 and M.Sc degree in Computer Engineering from National University of Sciences and Technology, Rawalpindi, Pakistan in 1996. He received his PhD degree from the University of Strathclyde, UK in Electrical Engineering with the thesis entitled, “Multimedia Communications over Mobile IP Wireless Networks”. Sajid has served as Research Fellow at University of Aberdeen from Jun 2012 to Mar 2015. He is currently a KTP Associate at London South Bank University, working as a Systems Engineer at Firstco Limited, London. His research interests are in adaptive video streaming, error resilience, video communications, and networking.

List of Abbreviations

4G	4th Generation
AF	Amplify-and-Forward
AL	Application Layer
AL-FEC	Application Layer Forward Error Correction
ARQ	Automatic Repeat Request
AVC	Advanced Video Coding
BER	Bit Error Rate
B-Frame	Bi-Directionally predicted Frame
BP	Belief Propagation
BPSK	Binary Phase Shift Keying
BS	Base Station
CIF	Common Interchange Format
CIP	Constrained Intra Prediction
DASH	Dynamic Adaptive Streaming over HTTP
DCT	Discrete Cosine Transform
DF	Decode-and-Forward
DP	Data Partitioning
DVB-H	Digital Video Broadcasting - Handheld
DVB-NGH	DVB - Next Generation Handheld
DVB-T	DVB - Terrestrial
DVB-T2	DVB-Second Generation Terrestrial
EEP	Equal Error Protection
ELP	Equal Loss Protection
ETSI	European Telecommunications Standards Institute
EW	Expanding Windows
EWF	Expanding Window Fountain Codes

EW-RLC	Expanding Window – Random Linear Codes
FEC	Forward Error Correction
FMO	Flexible Macroblock Ordering
FPS	Frame Per Second
FSMC	Finite State Markov Chain
GE	Gaussian Elimination
GF	Galois Field
GOF	Group of Frames
GOP	Group of Pictures
H.264/AVC	H.264/Advanced Video Coding
H.264/SVC	H.264/Scalable Video Coding
HARQ	Hybrid Automatic Repeat Request
HDTV	High Definition Television
HEVC	High Efficiency Video Coding
HPC	High Priority Class
HPL	High Priority Layer
HSDPA	High-Speed Packet Downlink Access
HTTP	Hypertext Transfer Protocol
IDR	Instantaneous Decoder Refresh
I-Frame	Intra Coded Frame
IP	Internet Protocol
ITU	International Telecommunication Union
LDPC	Low Density Parity Check
LPC	Low Priority Class
LPL	Low Priority Layer
LT	Luby Transform
LTE-A	Long Term Evolution- Advanced
MB	Macroblock
MBIU	Macroblock Intra Update
MBMS	Multimedia Broadcast Multicast Service

MC	Motion Compensation
MCP	Motion Compensated Prediction
MD	Multiple Description
MDC	Multiple Description Coding
ME	Motion Estimation
MPE	Multi Protocol Encapsulation
MPEG	Moving Picture Experts Group
NAL	Network Abstraction Layer
NC	Network Coding
NOW	Non-overlapping Windows
OFDMA	Orthogonal Frequency Division Multiple Access
PE	Probability of Error
PET	Priority Encoded Transmission
P-Frame	Predictively Coded Frame
PLP	Physical Layer Pipes
PLR	Packet Loss Rate
PS	Probability of Selection
PSNR	Peak Signal-to-Noise Ratio
QCIF	Quarter Common Intermediate Format
QoS	Quality of Service
RASSA	Rate Adaptive Selective Segment Assignment
RCPC	Rate Compatible Punctured Convolutional Codes
RGB	Red-Green-Blue
RLC	Random Linear Codes
RNG	Random Number Generator
RS	Reed Solomon
RTP	Real Time Transport Protocol
RTC	Raptor Code
SNR	Signal-to-Noise Ratio
SVC	Scalable Video Coding

TCP	Transmission Control Protocol
TDM	Time Division Multiplexing
UEP	Unequal Error Protection
ULP	Unequal Loss Protection
UDP	User Datagram Protocol
VCL	Video Coding Layer
VOD	Video-on-Demand
WiMAX	Worldwide Interoperability for Microwave Access
XOR	Exclusive OR
YUV	Brightness/Luminance/Chrominance Space

CHAPTER ONE

INTRODUCTION TO VIDEO COMMUNICATIONS AND STREAMING

1.1 Introduction

Video is the most important medium for communications. Video communication is unimaginable without compression due to the large size of video data. The larger video data file sizes necessitate a high degree of compression for video streaming data for its storage and transmission. Modern video coding standards employ motion-compensated prediction (MCP) between consecutive video frames and a high degree of compression. The video data also has delivery time constraints different to those of non-real time flows like file transfers. Video communication is now mostly over Internet protocol (IP) based packet networks. IP networks are best-effort and provide no guarantees of bandwidth, delay, and losses [1].

The compressed bitstream becomes highly susceptible to any data losses as even a small loss can result in losing synchronization between the sender and receiver resulting in error propagation to subsequent frames. Thus there is a need to devise schemes to offset the effect of data loss.

With video coding standards like H.264/AVC or SVC it is possible to encode the video stream so that its encoded content can be adapted to cope with the variations in bandwidth. Such adaptations are possible because of video scalability, that is, a subset of the encoded video data can be extracted which is a low quality representation of the complete data. H.264/SVC inherently provides scalability, that is, the encoded video data can be arranged in a hierarchy of importance starting at base layer and one or more enhancement layers. Similarly, H.264/AVC data can be prioritized into most important data (analogous to base layer), and other (one or more enhancement) layers of decreasing importance through data partitioning, slicing and other features. The most important encoded data is the base layer which can provide an acceptable quality. Any loss from the base layer will render the received video data useless. Each enhancement layer successively adds to the video quality, and with all the enhancement layers the best quality can be achieved. After encoding, it is possible to extract subsets of video data which can suit heterogeneous devices with different displays, memory and processing power. Although, it is possible to drop data from the enhancement layers but care must be taken as the enhancement layers also have dependencies. It can therefore be advantageous to design a drop policy taking into account the layer dependencies and to provide better loss protection for prioritised transmission to the important layers.

The rest of the Chapter is organised as follows. The video compression standards are discussed in Section 1.2. Section 1.3 covers the error correction. Multiple description coding is described in Section 1.4. IP and Network protocols are covered in Section 1.5 and 1.6. Multimedia communication standards are covered in Section 1.7. Section 1.8 and 1.9 briefly describe video streaming and some of its applications. Finally Section 1.10 concludes the chapter.

1.2 Video Compression Standards

Video is a succession of still images. An image represents light intensity in spatial coordinates most of which remains constant or varies little from one image to the next except in areas where there is motion or changes in ambient light.

The size of video data is many orders of magnitude greater than text or scalar data. Sophisticated video compression algorithms have been developed so that video data can be easily stored and communicated.

For colour images, a colour space conversion is first applied to convert the Red-Green-Blue (RGB) image into a YUV (brightness/luminance/chrominance space) where these components can be assigned different weights depending on the human visual perception, which is strongest for brightness.

Spatial redundancy is exploited by employing signal transformation, such as, Discrete Cosine Transform (DCT) which works on adjoining pixels and removes the redundancy. In order to remove temporal redundancy, the similarity between the frames which are closer together in time can be exploited. Instead of coding each frame in isolation, the similarity between frames is exploited by, first predicting it based on a previously coded frame, and then coding the difference in this prediction. In order to reduce the difference between frames, the motion is estimated by a process termed motion estimation (ME).

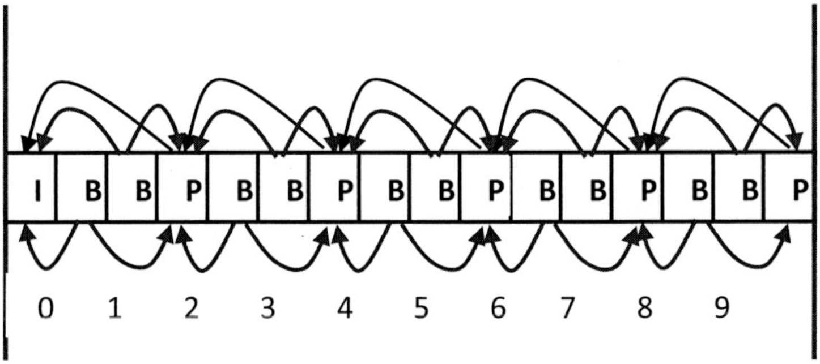

Figure 1-1: GOP structure of 16 frames.

A frame coded independently of other frames is termed as intra-frame or I-frame. Such frames typically exist at the start of video sequence or a group-of-pictures (GOP) and generally have larger sizes. A frame based on a prediction to a previous frame only is termed as a predictively-coded or P-frame. The prediction however could be done based on a previous and future frame as well, which gives even better compression, such frames are termed as bi-directionally predicted frames or B-frames. The different coded frames and their dependencies are shown in Figure 1-1. The selection of prediction dependencies between frames can have a significant effect on video streaming performance, e.g. in terms of compression efficiency and error resilience.

Video compression is normally lossy, that is after compression of the video data, its reconstruction back to the original may not be exact. However, such loss is acceptable and its nature may be governed by a given application. The latest video compression standards achieve compression by applying the same basic principles as described above. The colour space redundancy is exploited by a colour space conversion. Then the temporal redundancy is removed by applying Motion Compensation (MC)-prediction, and the spatial redundancy is removed by applying the DCT. The resulting DCT coefficients are subsequently processed to generate the compressed bit stream. However, as this compressed bit stream is highly susceptible to quality degradation even by bit losses hence the modern video coding standards employ quite advanced techniques to contain the adverse effect of such losses.

1.2.1 H.264/Advanced Video Coding (AVC)

H.264/AVC [2] is a recent state of the art video compression standard. Similar to the prior video coding standards, it is based on Motion Compensated Prediction (MCP), which requires maintaining synchronisation between the encoding and decoding operations.

H.264/AVC has achieved significant improvement in rate-distortion efficiency relative to existing standards [3]. H.264/AVC has been adopted by various application standards and is increasingly used in most video applications.

1.2.2 H.264/Scalable Video Coding (SVC)

SVC is an extension of H.264/AVC standard and adds scalability features to it. A scalable video encoder compresses a raw video sequence into multiple layers. Scalable video means that it is possible to extract a subset of compressed video data which can be used to playback the video at a lower quality. Scalability makes it possible to encode the video only once in different layers or classes, some of which (the least significant ones) can be removed to fit a bandwidth, processing power or display characteristics of the target device. Modern video transmission is over mobile heterogeneous devices and the video resolution and quality must adapt to the characteristics of the device. SVC is a highly attractive solution to the problems posed by the characteristics of modern video transmission systems [4].

The most important layer is the base layer, without which decoding fails but which provides a coarse quality at a much reduced data rate. Additional compressed layers are enhancement layers that provide additional quality to the received video stream. Enhancement layers can be decoded only in conjunction with the base layer. On a Quality of Service (QoS)-enabled IP network it would even be possible to send the base layer with a higher priority than the other layers.

The networks which support prioritization can make use of scalability by assigning a higher priority to the layers according to their importance.

The base layer could thus be transported with the highest priority. However, Internet does not provide any such prioritization and all packets are equally likely to be lost. In such networks, the scalability alone does not bring any advantage for video transport; however, channel coding can be used to make the base layer more tolerant to errors [6]. The layers can be combined to adapt to different frame rates, spatial resolutions, or quality, of the video content, giving rise to temporal, spatial, and quality scalability.

1.2.3 High Efficiency Video Coding (HEVC)

HEVC is the latest video coding standard which provides 50% reduction in bandwidth for the same quality as compared to H.264 standard. The high coding efficiency will enable video communication over low-power and bandwidth-constrained mobile devices.

HEVC is block based and utilises many of the coding concepts from H.264/AVC standard in innovative ways to achieve a high degree of compression.

1.3 Error Correction

The Internet is prone to packet losses which can result from packet drops due to congestion or bit errors. Such losses can degrade the user experience and therefore measures must be taken to recover from losses and provide a graceful degradation in video quality.

There are different means available to protect the video data and to increase the quality of the user experience at the receiver. This could be accomplished by providing:

- Error resilience
- Error protection
- Error concealment

The data transmission based on IP is in the form of packets where each packet may be independently routed. Some of the packets may be lost en-route e.g., due to buffer overflows or may be delayed beyond their display deadline. The traditional solutions for data delivery with re-transmission of lost packets as in Transmission control protocol (TCP) does not work well for real-time video transmission because of the tight delay constraints of each packet. As TCP is based on an acknowledgement of packets from the receiver so it will give poor performance for transmission over large distance. One of the solutions is to use channel coding techniques which could recover the original data despite losses. The solutions based on schemes like Reed Solomon (RS) codes are inflexible because the code rate has to be fixed in advance. Moreover, the encoding and decoding operations are quite complex especially for large Galois Field. For such codes the error characteristics of the channel must be known in advance in order to adjust the code rate to it. This solution does not extend well to multiple receivers as then only a worst-case erasure channel can be assumed for all receivers.

1.3.1 Error Resilience Features

The latest video standards are aimed at packet-based networks and thus cope mainly with packet losses instead of bit errors [2]. There are many error-resilience features built into the latest coding standard aimed at curtailing the video degradation at the receiver if some information is lost during transmission.

Figure 1-2: Effect of a lost slice is restricted to smaller area.

1.3.1.1 Slice Structured Coding

A frame may be divided into suitably sized portions or slices to provide protection against degrading the whole frame during transmission. Each picture may be split into one or several slices as shown in Figure 1-2. The macroblocks will be organized in slices and this has the desirable effect of making transmitted packets smaller; making the overall recovered video frames better tolerant of errors. Slice structured coding also introduces slice headers to act as resynchronisation points to localize the errors and prevent error propagation to subsequent frames. Each slice can be correctly decoded without the use of data from other slices provided in the same frame. Each slice is encapsulated in a separate packet by H264/AVC encoder. Slices are self-contained, thus, if a coded slice is available to the decoder, all the MBs in that slice can be decoded.

1.3.1.2 Data Partitioning

DP is a feature available in the extended profile which supports the partitioning of a frame/slice in up to three partitions, based on the importance of the encoded video syntax elements for video reconstruction [92], [93].

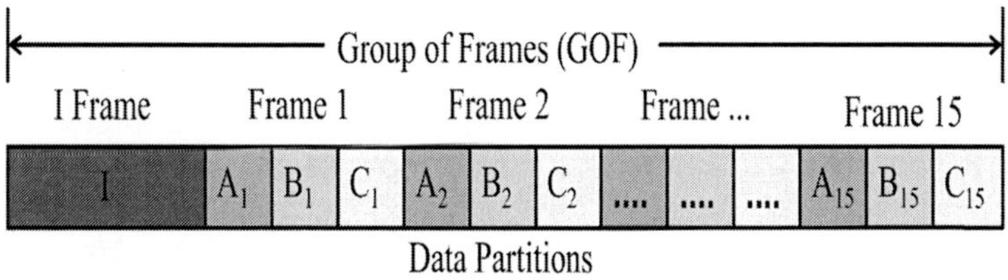

Figure 1-3: Relative position of DP A, B and C in a GOF of 16 frames.

A GOF with each frame split in three partitions is shown in Figure 1-3. DP A contains the most important data comprising slice header, quantization parameters, and motion vectors. DP B contains the intra-coded macroblocks (MB) residual data, and DP C contains inter-coded MB residual data [92], [89] © 2012 IEEE, [94] © 2013 IEEE. The decoding of DP A is always independent of DP B and C [91], [90] © 2011 IEEE. At the decoder, if the Type B or Type C partition is missing, the header information in the Type A partition may be used in order to aid in error concealment.

However, if DP A is lost the remaining partitions cannot be utilized. The decoding of DP B is possible without DP C, but not other way around. To make DP B independent of DP C, Constrained Intra Prediction (CIP) parameter in H.264/AVC encoder must be set [51], [91], [92], [93], [58], [90] © 2011 IEEE, [89] © 2012 IEEE.

1.3.2 Fountain Codes

Fountain codes are rateless codes which can potentially generate an unlimited number of encoded packets from a limited number of source packets. Each encoded packet is based on a combination of the source packets according to some distribution. It is thus not necessary as to which packets are received but rather that they are received in sufficient quantity. The decoding is possible if the number of received encoded packets at the receiver is just a little more than the source packets. The encoding thus eliminates the effect of independent losses at different receivers, and also there is little requirement to send feedback to the sender.

For the application of Fountain codes to video data, a Group of pictures (GOP) could be treated as a source block. The encoded packets for a particular GOP are generated based on the available bandwidth. The solutions based on Fountain codes adapt well to the varying bandwidth as the receiver or sender can terminate reception/transmission depending on the available bandwidth. Such codes are ideally suited for use in multicast scenarios because there is no requirement to target a particular receiver. It is thus imperative to design Forward Error Correction (FEC) solutions which are adaptive to the varying wireless channel conditions, i.e., the bandwidth and the packet loss rates. This adaptive behaviour becomes even more important for heterogeneous receivers.

The FEC protection could be provided at different layers of the network protocol stack. However, providing the FEC solution at the application layer (AL) makes it more flexible. It can also be easily implemented in software. In addition, a video application knows best how to handle each packet and therefore it is better to leave such decisions to be taken at the application layer. Error resilience features and FEC at the AL are thus often used to protect the data against losses.

The Third Generation Partnership Project (3GPP) recommends the use of FEC for Multimedia Broadcast and Multicast Services (MBMS) and, more specifically, adopts the use of Raptor FEC code in the AL. Digital Video Broadcasting-Handheld (DVB-H) uses Raptor codes and similar AL-FEC schemes are proposed for DVB-Next Generation Handheld (NGH) and Long Term Evolution-Advanced (LTE-A) at the AL. Due to backward compatibility with DVB-Terrestrial (T), DVB-H uses the same first generation FEC schemes with convolutional coding and Reed-Solomon (RS) block coding in addition to AL Raptor codes.

Other than Raptor codes, a class of rateless codes that is gaining increased popularity for applications in wireless broadcast/cellular networks are Random Linear Codes (RLC) [68] © 2011 IEEE. With multi-hop and cooperative communications becoming popular in emerging wireless network architectures, introduction of AL RLC may serve as a step forward towards exploring the benefits of network coding. In addition, the emerging networking concepts, such as hybrid broadcast/cellular networks (with users equipped with multiple interfaces) or device-to-device communications, offer a number of opportunities for achieving network coding gains using RLC [68] © 2012 IEEE.

1.3.3 Unequal Error Protection

The amount of redundancy added by AL-FEC can be adapted to provide better protection where it is most needed. The whole of the encoded video data is not equally important for the video re-construction, rather different importance classes can be identified within it. The most important data for video reconstruction is termed as base layer and provides an acceptable quality. Any error in the base layer causes severe degradation. The remaining data resides in one or more enhancement layers which progressively improve the quality of video. The enhancement layer is useless by itself and it is required that the base layer is also received for it to be of some benefit. Therefore it is important to provide more protection to the base layer. Such prioritized protection of the data is termed as unequal error protection (UEP). It is thus advantageous to exploit this aspect while designing FEC solutions.

1.4 Multiple Description Coding

Multiple Description Coding (MDC) splits a single media stream into n sub streams or descriptions. These descriptions can then be routed over multiple paths to a destination. The paths may be partially or fully disjoint which ensures that even with adverse channel conditions on one path, some descriptions will make it through to the destination. It should be possible to recreate a low quality media stream using any one description. However, with reception of more descriptions the quality of media stream increases progressively. The idea of MDC is thus to provide error robustness to the transmitted data.

A very simplified way to create two descriptions could be to partition a video stream into odd and even frames and encode them separately, which could later be sent on different paths. With multiple descriptions, correct reception of just one description can results in an acceptable quality. With more descriptions received correctly, better quality can be obtained [6].

Coding into multiple descriptions is accompanied by a loss in compression efficiency. MDC can thus be applied successfully if loss in compression efficiency is balanced by the gain through overcoming transport losses [30].

1.5 Internet Protocol (IP) for Everything

Internet has enabled the digital technology revolution. Its success is derived from its flexibility in supporting different protocol layers for any type of physical media. IP provides a mechanism for transporting packets between devices connected over the Internet; however, it was never designed for video data. Its connectivity around the globe has led to more and more services being offered over the IP including video streaming besides others. Coupled with the improvements in the network infrastructure this trend is likely to increase with time.

The convergence of traditional television broadcasting and cellular communication is taking place. The mobile receivers are now capable of broadcast reception and transmission in addition to the other wireless communication capabilities like Bluetooth, wireless Local Area Network (LAN), and cellular communication.

Depending on the relationship between the sender and number of receivers, the communication can be classified as unicast, multicast or broadcast. In a unicast transmission one sender is connected to one receiver. The main advantage of such a system is that the transmission can be tailored to the receiver based on the feedback from it. In multicast, one sender serves multiple receivers. Although multicast is efficient in terms of resource utilization, it lacks the ability to target the transmission to a particular receiver. In broadcast a sender transmits to all the reachable receivers in the network.

1.6 Network Protocols

Transport protocols are used in conjunction with IP to define the transmission of packets. IP is the network protocol used for most communications applications and this trend is likely to grow. The data is sent as series of small packets over the network.

1.6.1 Transmission Control Protocol (TCP)

TCP is a complex protocol as it provides a guaranteed delivery through timeouts and retransmits. It provides a reliable delivery service by establishing a connection between sender and receiver before any data transfer can take place. It keeps track of the data transferred and retransmits the lost data through acknowledgements of the received data. Thus it can handle lost packets and provides flow control so that a sender only transmits at the rate at which data can be consumed at the receiver. This is not suitable for video data transport as retransmissions of lost packets may arrive after their decoding/display deadline.

1.6.2 User Datagram Protocol (UDP)

UDP is a simple protocol compared to TCP. It does not have the connection establishment and does not provide a guaranteed delivery, hence does not have acknowledgements and flow control. Although UDP seems to be a natural choice for video data but it requires the error detection and flow control to be provided at the application layer. The sender can control how fast the data is injected into the network but this could cause problems for other flows competing with a video flow over UDP.

1.6.3 Real-time Transport Protocol (RTP)

RTP or Real-time Transport Protocol and RTCP (Real-time Control Protocol) were designed to provide support for real time multimedia communications. RTP adds functionality for video transport over UDP without the unnecessary functionality of TCP.

1.6.4 HTTP Streaming

There has been a recent interest in streaming over standard HTTP protocol. This is a refinement over progressive download and the playback starts immediately. The video data is downloaded in small segments and played back at the same time. It is done over TCP but takes the advantage of Internet's high speed infrastructure.

1.7 Multimedia Communication Standards

1.7.1 Digital Video Broadcasting

DVB-H is the digital broadcast standard for the transmission of broadcast content to handheld terminal devices [23]. DVB-H is based on the DVB-T standard for digital terrestrial television for the mobile class of receivers.

DVB-H is the European standard for digital TV signals broadcast to handheld devices and is based on the DVB-T standard which mostly uses MPEG-2 packet streams. DVB-H is IP-based and has two important new features introduced in it. Firstly, the transmission takes place in intermittent bursts of maximum data rate of up to 2 Mbps. Secondly, to mitigate errors in mobile and wireless environments the DVB-H standard introduces an optional link layer FEC mechanism, called MPE-FEC [35], [55], [58], [90] © 2011 IEEE.

The experience in establishing the DVB-H services have shown that there is a relationship between DVB-T and DVB-H services [24]. With the introduction of DVB-T2, there is a need to look at ways of leveraging the advantages of DVB-T2 in developing NGH in the new paradigm of rich media content delivery in the convergent era. DVB-NGH will be based on DVB-T2 and as DVB-H uses DVB-T physical layer, the physical layer of DVB-NGH will be DVB-T2 based [25].

As DVB-H shares the same physical layer of DVB-T this leads to some restrictions which could be overcome by a better Time Division Multiplexing (TDM) -system design [28] enabling a more flexible system. Also, in DVB, power saving efficiency and the burst size are linked together in a delicate manner which requires careful tuning of the parameters, which would be done in an independent manner in DVB-NGH.

Assuming that the H.264/AVC standard is used for video encoding, DVB-H protocol stack successively adds headers and prepares the video data for transmission. NAL video packets of H.264/AVC are encapsulated into RTP/UDP packets and placed into IP packets, which are inserted into MPE sections, sorted into MPE frame and divided into transport stream (TS) packets for transmission over the DVB-H physical layer. [55], [68] © 2011 IEEE. All TS packets belonging to one MPE frame are sent in one transmission burst. Note that, if AL-FEC is used, the link-layer FEC mechanism (based on RS codes) is switched off [91].

In general, one IP packet can be contained in multiple TS packets. Moreover, a single TS packet can contain data from two consecutive IP packets. Clearly, with each encapsulation the additional header content is appended making the header overhead substantial especially for small size IP packets. Therefore, the performance of AL-FEC codes depends not only on the erasure decoding performance of the FEC scheme, but also on the flexibility of the FEC scheme to provide different provide FEC configurations which can be adapted to the specific protocol structure at lower layers [55] © 2011 IEEE.

1.7.2 Long Term Evolution- Advanced (LTE-A)

LTE is an Orthogonal Frequency Division Multiple Access (OFDMA)-based, mobile broadband technology developed by The Third Generation Partnership Project (3GPP). 3GPP initiated its LTE standardization work at the end of 2004, to meet the ever increasing requirements of better QoS and higher wireless data rate. This standardization was successfully completed by the end of 2007 [26]. Thereafter work was started on LTE-Advanced standardization process to address the requirements of Fourth Generation (4G) IMT-Advanced systems by considering new transmission technologies [27], like adaptive interference management, coordinated multiple point transmission and reception, and relay.

The next significant performance leap will come from heterogeneous networks, which bring the network closer to the user through small cells such as picocells and femtocells. This ensures an enhanced mobile experience with higher data rates to more users. The data rate between the terminal and the base station is dependent on many factors. [29].

1.8 The Video Streaming Revolution

Video streaming is generally based on a client/server model. A client requests the video from a server which delivers the data to be played by the client. A client can start to play the video data before the entire file has been received from the server.

Live streaming refers to the case where the video data does not exist in stored form (Video on demand) but rather is being captured by a camera as it happens, compressed and transmitted to the receiver. Live video streaming thus poses more challenges and constraints as compared to video on demand.

The advances in communication infrastructures such as the Internet and wireless mobile networks are the enabling technologies for the widespread use of multimedia applications. In terms of applications, these range from a simple music download to watching TV over mobile devices. Some of these applications over Internet protocol (IP) are traditional, such as voice-over-IP (VoIP) to a conventional telephone, or sending television programs over IP to a conventional set top box.

Multimedia traffic over IP based wireless networks has been increasing steadily over the recent past. With further developments in multimedia services and communication technologies, it is projected that more and more services will be interactive. Multimedia communications over best-effort packet networks such as Internet is quite challenging because of the dynamic and unpredictable delay, loss rate, and available bandwidth. Video transmission therefore must adapt to the varying channel conditions.

1.9 Applications of Video Streaming

Video streaming has become ever more prevalent with the increase in network bandwidth to support the real-time video traffic. The video streaming applications have emerged in the following application areas:

1.9.1 Medical

The video streaming can be used to aid the remote medical diagnosis and surgeries by the specialists who can guide the on-site medical teams by providing guidance through access to the remote live video stream.

1.9.2 Security and Surveillance

The domain of security and surveillance requires live or near real-time video data for viewing and recording. This enables the security agencies to take the preventive and proactive actions.

1.9.3 Education

The video streams can be used to impart lectures and training independently of the time and space constraints. Pre-recorded content can be covered at a pace as determined by the learner.

1.9.4 Entertainment

The live streaming of videos over Internet has become a big industry. Anyone can upload a video on any topic while those interested can watch the video content of their choice.

1.9.5 Interactive Computer Games

The gaming industry depends heavily on the video streaming with stringent timing deadlines to maintain interactivity between multiple players.

1.9.6 Television Broadcast

The television is gradually morphing into an interactive application where the viewers are not mere captive audiences of the broadcaster but can also interact and personalise the viewing experience.

1.9.7 Videoconferencing

The videoconferencing diminishes the distance barrier and people can interact naturally. This application has tight deadlines to maintain interactivity and obviates the need for travel saving time and cost to the parties involved.

1.9.8 Sports

The live streaming of sports events has grown into a big industry and this trend is likely to grow further in view of the large followership of games and sports. With the technology advancements more and more visual information captured from various angles is available to the viewers.

1.9.9 Remote Monitoring and Surveillance

The monitoring of remote sites has been enabled by miniaturization of electronics and embedded applications. This monitoring could be for critical infrastructures, asset management and security.

1.10 Conclusion

This chapter provides the necessary background for the following chapters. A discussion of the latest video compression standards is provided. The error-resilience features which are used in later chapters to design novel solutions are described.

A section on MDC describes the basic principles and architecture of the technique. Network protocols and multimedia communication standards have been introduced. These topics are used in later chapters along with their application to specific research problems.

The video streaming has started to play a major role in the emerging and innovative applications and this trend is likely to grow further. The increase in scale and sophistication of video streaming will bring new challenges which would need combined application of video coding algorithms, better error resilience techniques, and exploitation of error correction codes with unequal loss protection.

CHAPTER TWO

FOUNTAIN CODES FOR ERROR CORRECTION

2.1 Introduction

The video data contains elements, e.g. slices, DPs that have different importance for video reconstruction. It is thus beneficial to group the data based on its importance, and then provide different degree of protection using UEP FEC, with the important data having more protection. DP feature can easily partition H.264/AVC video data into different priority classes. The proposed schemes exploit the DP H.264/AVC video transmission using Raptor and RLC, and their performance is investigated as AL-FEC solutions in the DVB-H and DVB-T2 networks. For the analysis, DVB-T2 network is configured according to the DVB-T2 Lite profile [31], which is mainly intended for mobile reception. The RLC results are extended to provide comparisons between NOW RLC and EW RLC, which are two effective UEP RLC strategies. The results obtained using realistic simulated DVB-H and DVB-T2 channel traces show viability of the EW RLC as a promising AL-FEC solution for multimedia broadcast applications due to its robustness to varying channel conditions.

Using FEC mechanisms is the most favoured approach as retransmissions in broadcasting applications are usually counter-productive [32], [33], [34]. One of the major arenas where FEC codes are successfully applied for wireless streaming video protection in recent years are DVB networks, in particular, their extensions for handheld devices called DVB-H and next generation handheld (DVB-NGH) which will be based on DVB-T2. One of the video coding standards approved for DVB-H and DVB-NGH is H.264 Advanced Video Coding (AVC) [2].

For error protection of broadcasted video, at the link layer, DVB-H specifies a novel Multi-Protocol Encapsulation - Forward Error Correction (MPE-FEC) solution [23] for real-time services with adaptive punctured Reed-Solomon (RS) protection against packet losses. The DVB-H standard provides a possibility of using Digital Fountain Raptor codes as the AL FEC solution for IP datacasting services [9], [33], [35]. Though

Raptor codes are currently proposed only for non real-time services, it is expected that emerging DVB-NGH will enable the use of AL-FEC solutions for streaming applications as well [90] © 2011 IEEE. Motivated by the need to provide better high definition television (HDTV) services, DVB-T2 brings higher throughput than DVB-T.

Apart from Raptor codes, a class of rateless codes that have been gaining increased popularity recently for applications in wireless broadcast/cellular networks are RLC [13], [36]. In addition and for possible future extensions, as multi-hop and cooperative communications are becoming increasingly popular in emerging wireless network architectures, introduction of RLC may serve as a step forward towards exploring the benefits of network coding [37]. In the context of future DVB networks, this may be important for emerging concepts such as hybrid broadcast/cellular networks (with clients equipped with multiple wireless broadband interfaces) and device-to-device communications [38], [39], [91], [58][90] © 2011 IEEE.

The focus of this chapter is to analyse the use of the DP error resilience feature of H.264/AVC in combination with Raptor codes and RLC as an AL-FEC solution for video transmission in mobile DVB networks. Furthermore, the benefits of UEP RLC as an AL-FEC solution, tailored with the DP feature of H.264/AVC for video transmission in the DVB-H standard, is investigated in detail. The DP feature of H.264/AVC enables simple rate adaptation crucial in wireless broadcasting [23], [40], [41]. With DP the percentage of entire frames being lost can be reduced [42]. The use of slice partitioned H.264/AVC with UEP RLC is also covered. An alternative is Scalable Video Coding (SVC) [4] possibly together with some type of error resilience [40]. Wireless video broadcasting of H.264/SVC, in combination with rateless codes, has been studied extensively (see [33], [40], [43] and references therein). The main advantage of DP H.264/AVC over SVC is its compatibility with H.264/AVC. The UEP using DP of H.264/AVC is proposed in [5]. DP H.264/AVC and UEP are later used for wireless video delivery in [44] using rate compatible punctured convolutional (RCPC) codes, in [45] with hierarchical quadrature amplitude modulation, in [46] with Raptor codes for IPTV, and in [47] with Growth codes. In [48], design of a Raptor generator matrix based on frame dependencies within a group of picture (GOP) has been proposed. The redundancy allocation process is tied to the knowledge of the channel loss which may not be suitable for real-time applications and video broadcasting. The proposed scheme also needs a mechanism to transport the generator matrix coefficients to decoder [91].

In [49], the Expanding Window Fountain (EWF) codes, as LT [8] UEP codes, are proposed and their asymptotic behavior analytically found; EWF is a layered scheme where the protection of lesser important layers also includes the more important layers. In [21], EWF codes are optimized for scalable video delivery. In [34], a layer-aware FEC mechanism has been proposed, similar to the EWF concept. However, in contrast to [21], [49] in [34] UEP low-density parity-check (LDPC) codes and Raptor codes are designed for protection of H.264/SVC video. The work also explores the suitability of the proposed codes to physical and application layer FEC protection for the latest video communication standards including DVB-H. Extending the UEP design methods of [21] from Raptor codes to RLC, UEP-based RLC strategies for the SVC delivery have been investigated in [14], [22] and [50].

While [22], [50] are more focused on performance analysis of UEP RLCs schemes over random erasure channel and SVC [4], this chapter addresses design challenges of UEP RLCs schemes with DP feature of H.264/AVC for real-world implementation in the mobile DVB networks and compares their performance to standard Raptor codes. While both Raptor codes and RLC have been individually well studied in literature, interestingly, no systematic performance/complexity comparison has been conducted. The key contribution of the chapter is comparison between Raptor and RLC codes for DVB-H and optimal code design guidelines for UEP RLC over DVBH and DVB-T2 configured according to the DVB-T2 Lite profile, which is a profile intended to allow simpler receiver implementations for very low capacity applications such as mobile broadcasting, although it may also be received by conventional stationary receivers [91]. Thus, this chapter evaluates performance of the proposed schemes in a real-world environment and as a result highlights set of optimized design recommendations.

This chapter is organized as follows. The fountain codes are described in Section 2.2. The unequal loss protection schemes are covered in Section 2.3. The proposed video broadcasting system is described in Section 2.4. The video broadcasting with Raptor and RLC is discussed in Section 2.5. Section 2.6 discusses the UEP AL-FEC for video broadcasting and performance evaluation of EW-RLC based solution with DVB-T2. The slice partitioned H.264/AVC video broadcasting with UEP RLC is covered in Section 2.7. Finally, Section 2.8 provides the conclusion.

2.2 Fountain codes

Fountain codes are erasure correction code, capable of correcting lost data. This ability to recover the lost data without requiring retransmission provides reliability and efficiency in data networks. These codes have transformed the conventional paradigm of ordered delivery for transmitting the packetized data over erasure channels.

With fountain codes approach a user can receive any packets (a little more than the source data) in any order to be able to decode the given data. In case of transmission to large number of users, it results in a much simplified data delivery [7].

Fountain codes are rateless codes, that is, given k packets of a file, potentially infinite encoded packets can be generated. If any k $(1 + \alpha)$ packets (irrespective of the order) are received, the receiver can reconstruct the original data. α is a non-negative small fraction less than 1, and is very small for large k. Fountain codes are application layer- forward error correction codes generally implemented in software, to protect packet losses.

A systematic code is any error-correcting code in which the input data is embedded in the encoded symbol. The advantage of such codes is that the receiver does not need to recover the original source symbol in case of correct reception. Also, systematic codes reduce the decoding complexity since with systematic codes the decoder operates with the matrix that has reduced number of rows (reduced, by the number of correctly received systematic packets). However, in general systematic codes do not provide improvement compared to non-systematic codes.

2.2.1 LT Codes

Luby transform (LT) codes are the first class of fountain codes [8]. It works on a very simple algorithm based on exclusive OR (XOR) operation to encode and decode the message. The LT encoder can generate a potentially infinite sequence of encoded symbols from a set of k source symbols.

These encoded symbols can then be injected into the channel. Some of the symbols will be lost and assume that the symbols correctly received at the decoder is little more than k. Before the decoder can attempt to recover the original source symbols it needs to know the degree of each symbol and the connected symbols (exclusive-OR). There could be a variety of ways to pass this information to the decoder and a simple one is to pass a random number generator (RNG) seed along with each encoded symbol.

The degree distribution used in the encoding process is a critical part of the design. Some of the encoded symbols should have enough high degrees to avoid the case where a source symbol is not connected to any encoded symbol at all. On the other hand, some of the encoded symbols should have lower degrees to start the decoding process. During the decoding process, ideally, at any time, there should be only one encoded symbol with degree one and after processing this degree-one encoded symbol, the degrees of the remaining encoded symbols are reduced in such a way that only one encoded symbol has degree one. In [8], the ideal Soliton distribution is specified. Also, a refinement over it termed as Robust Soliton distribution which provides better results is proposed.

2.2.2 Raptor Codes

Raptor codes as initially described in [9] extend the idea of LT codes one important step further. Raptor codes use two encoding stages for encoding, consisting of a pre-code followed by reduced-complexity LT coding. Raptor codes achieve linear time encoding and decoding by pre-coding of the input symbols before the application of LT code.

Raptor codes [9] are rateless codes, i.e., they can generate as many encoded symbols as desired from a given set of source symbols. The Raptor decoder can recover the original source symbols from any set of encoded symbols, as long as their number is at least equal or slightly exceeds the number of source symbols [91], [58] © 2011 IEEE.

These have recently been adopted for use in various standards including DVB-H. As an AL-FEC solution in DVB-H, systematic Raptor codes provide improved system reliability and a large degree of freedom in the choice of transmission parameters [11], [91], [58] © 2011 IEEE.

The paper by Shokrollahi [9] has many additional details, including useful descriptions of extremely effective practical constructions and analysis techniques for codes of finite length. Raptor codes currently give the best approximation to a digital fountain. A virtually limitless supply of packets can be generated on the fly after some small initial pre-processing, with each packet taking only constant time to produce. Decoding can be accomplished after receiving just a few percent more than the minimum of k encoding packets (with high probability), and requires space and time linear in the size of the original message. Moreover, very efficient implementations are possible.

For an explanation of the encoding and decoding algorithms and implementation guidelines, [12] is a detailed reference. The decoding process results in maximum-likelihood decoding performance and it can be considered successful if the received generator matrix at the decoder is invertible [55] © 2011 IEEE.

2.2.3 Random Linear Codes (RLC)

A class of rateless codes which has become popular recently are RLC [13]. RLC applied over a source message produces encoded symbols as random linear combinations of source symbols with coefficients randomly selected from a given finite field [10], [93], [55][68][90] © 2011 IEEE.

As a packet- level AL FEC solution, RLC is simple to implement and perform as near-optimal erasure codes [12][14] for sufficiently large finite field used for creating linear combinations of source symbols [55] © 2011 IEEE (one-byte Galois field GF(256) is usually sufficiently good). This makes RLC an attractive alternative to Raptor codes as a universal FEC/network coding solution for emerging wireless communication systems, such as LTE-A, Worldwide Interoperability for Microwave Access (WiMAX), and DVB-NGH [15][16][17][18][19][91][93], [58][90] © 2011 IEEE.

If there are k source packets to be transmitted, then for each encoded packet, the source packets to be exclusive-OR are determined. That is each encoded packet is an exclusive-OR (in the selected field) of the selected source packets. Some of these encoded packets will be lost in the channel. The decoder receives the remaining packets which must be at least equal to k for the decoding to succeed. The decoder must be able to create the generator matrix in order to proceed with the decoding. Several methods could be used for this.

If the generator matrix at the decoder is invertible then the decoder can recover the original k source symbols. By performing Gaussian Elimination (GE) the inverse of the matrix can be computed. The major limitation of the application of RLC is the decoding complexity of GE decoding, which is polynomial in the number of symbols. However, for short lengths of the source messages, the decoding complexity is acceptable [55], [58] © 2011 IEEE (also see [20] and references therein). Moreover, there is no performance penalty for using short-length codes as compared to standard rateless codes [91], [90] © 2011 IEEE.

2.3 Unequal Loss Protection

The elements of compressed video data do not have same contribution for the video re-construction quality. It could thus be beneficial to design effective error protection schemes which exploit this inherent prioritization of video data for transmission. A scheme which treats all the data to be transmitted with Equal Error Protection (EEP) does not take into account the different priorities of video data into account for providing error protection. On the other hand, protection schemes can be designed which assign a degree of protection to a class of video data in accordance with its importance. Such a protection scheme is termed UEP. In order to define a class/window over a subset of source symbols, all the high-priority data units belonging to a particular priority have to be aggregated together over the whole source data. The classification of video data could be based on DP, slicing, type of frames, or any other criteria.

The rateless codes can be used to provide UEP. The basic idea is to design the channel codes so that the more important data for video re-construction is better protected. Generally, two or three windows are deemed sufficient.

2.3.1 Non-overlapping Windows (NOW)

A simple UEP scheme could be designed by dividing the source data into non-overlapping sets, termed "windows", which represent different classification of video data in accordance with its importance for video re-construction [50].

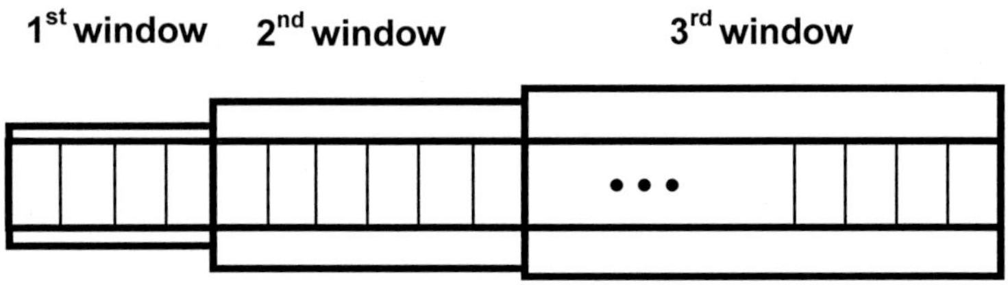

Figure 2-1: Non-overlapping window structure.

A general layout of the window structure with three priority classes is as shown in Figure 2-1. The most important data will reside in window W1, and the subsequent windows will hold data with a progressively decreasing priority. In order to generate channel symbols, first a window has to be selected whose source symbols will be used. Therefore, in order to provide UEP all that is necessary is to assign a Probability of Selection (PS) to each of the windows. The window with most important data could be assigned a higher PS relative to the other windows. After a window has been selected then the coding procedure is same as described in Section 2.2 for the various codes.

With this layout of windows, the source symbols from each window are encoded and decoded independently of the other window. In order to decode the whole of video data in all windows, all the decoders must succeed. It is possible to alter the window structure such that the low-priority windows also aid the decoding process of the high-priority windows. Such a window structure is termed as Expanding Window (EW) [50].

2.3.2 Expanding Windows (EW)

Expanding window fountain (EWF) codes [21] are a class of UEP fountain codes based on the idea of creating a set of "nested windows" over the source block. The rateless encoding process is then adapted to use this windowing information while producing encoded packets. To obtain source blocks amenable to UEP, the set of windows is defined over the groups of source symbols of unequal importance. The coding is then performed over progressively increasing source block subset windows aligned with this "most to least importance" subsets. The general layout of a window structure with three importance classes is shown in Figure 2-2 [68], [90] © 2011 IEEE.

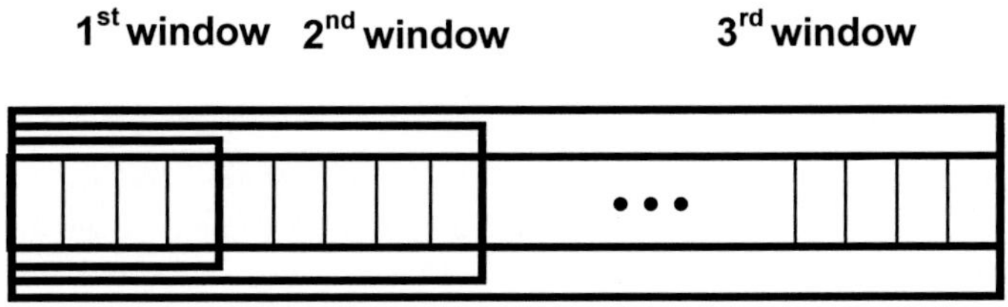

Figure 2-2: Expanding window structure [90] © 2011 IEEE, [89] © 2012 IEEE, [94] © 2013 IEEE.

The window with the most important subset of encoded data is the first window (W1) and the importance of data additionally included in windows progressively decreases on moving to the third window (W3). The subset data of W1 is contained in all the subsequent windows and is hence the best protected. Apart from W1, each window in addition to some of its own data also encloses the data of the higher importance windows [68] © 2011 IEEE. The size and structure of a window depends upon the elements meeting particular set criteria from a specific subset window. The number of windows is governed by the aggregation scheme employed to group encoded elements [93], [90] © 2011 IEEE, [89] © 2012 IEEE.

As in the case of NOW, the encoding process for EW has one important initial step that is to first select a window from which the encoded symbol is to be generated. This selection of a window is determined by Probability of Selection (PS) of a window which is a pre assigned parameter. PS is determined by the importance of different layers and the data rate available. After a window is selected, the encoding is the standard encoding performed over the source packets contained in that particular window only [22], [68][90] © 2011 IEEE.

The decoding of a window is same as standard decoding, in that, a window is recoverable if the receiver collects at least the same amount of linearly independent encoded symbols obtained from the window (or the windows contained in it) as there were in the window [22], [90] © 2011 IEEE, [89] © 2012 IEEE.

2.4 Proposed Video Broadcasting Based on DP H.264/AVC

In the following, a source-channel coding scheme is proposed for video broadcasting over DVB-H that exploits both state-of-the-art error resilient video coding and rateless AL-FEC to adaptively and optimally protect video against packet losses [91]. The block diagram of the proposed system is shown in Figure 2-3.

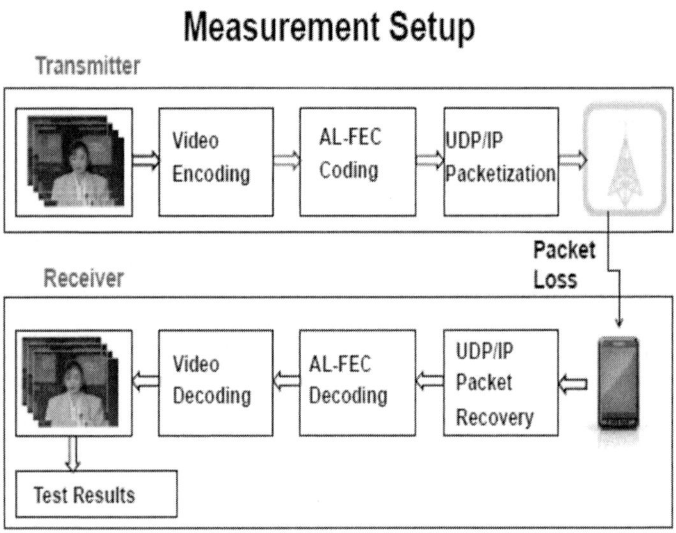

Figure 2-3: Block Diagram of the Proposed System.

2.4.1 Video Encoding

For each non-I frame, the video encoder output is split into three partitions: DP A, DP B, and DP C. The encoding is done with the Constrained Intra Prediction (CIP) flag set and the Macroblock Intra Update (MBIU) feature to limit the effect of error propagation. The cost of using CIP [51] can be neglected. All the partitions of a particular type from all frames within one GOP are extracted and aggregated together. That is, DPs A from all encoded frames of a GOP together with the encoded I frame are grouped generating the most important source class or layer. DPs B and C form second and third importance class/layers, respectively [91].

Each of the layers is packetized into equal-length source symbols/packets. Note that the length of the source symbols determines the number of source symbols contained in the source message, *k*, which directly influences FEC efficiency and decoding complexity. For Raptor codes, ideally, short source symbols should be selected resulting in higher values of source message length *k*, typically in the order of thousands of symbols, since decoding complexity increases only linearly with *k*, and on the other hand, performance is improved for larger *k*. In the case of RLC, however, large source symbols are favoured in order to yield small values of *k*, typically in the order of hundreds, because of high decoding complexity of RLC [58] © 2011 IEEE. The key issue to be investigated is the impact of different selection of *k*, imposed by performance and complexity issues, to these two codes [91].

2.4.2 AL-FEC Coding

A straightforward approach to AL-FEC is to apply one code over the entire source block (i.e., a GOP), that is, to equally protect the whole stream. This, equal error protection (EEP), however, does not capture adequately unequal importance of the three classes of the DP H.264/AVC code stream [91].

A preferable option is a UEP scheme which assigns redundancy to the source classes based on their importance to video reconstruction. Two different UEP design approaches are explored. The first one, NOW scheme uses a separate AL-FEC code, either Raptor or RLC, for each of the DP layers, and by using the codes of different code rate, UEP is realized. The second approach uses the expanding window approach [21], [22] to encode jointly all the classes [91].

The encoding process for both NOW and EW codes has one important initial step, that is, to first select a window from which the encoded symbol is to be generated. This selection of a window is determined by selection probability of a window which is a pre-assigned parameter keeping in mind the importance of different layers and the data rate available [91].

2.4.3 Data Packetization

After AL-FEC coding, NOW or EW with Raptor or RLC codes, encoded symbols are grouped to fit the content of a single IP packet. This process also takes into consideration the header overhead. IP packets are placed into the MPE frames where each IP packet is encapsulated within a single MPE section with its own overhead containing the error-detection field [91].

An MPE frame is transmitted within a single transmission burst by mapping the MPE frame data onto 188-bytes long physical-layer TS packets. The IP packet size is set to be an integer multiple of 184 bytes (the content size of TS packet), i.e., more precisely, each MPE section is transmitted over the integer number of TS packets [55], [58] © 2011 IEEE.

It is important to underline that IP packets are either correctly received at the decoder or lost in the transmission process. An IP packet is considered lost, if at least one TS packet that forms it is lost, which is the standard operation of the DVB-H link layer ([11], [52]) [91].

2.5 H.264/AVC Video Broadcasting with Raptor and RLC

In this section, we experimentally compare the performance of the standard Raptor codes with RLC for different source message sizes in order to select source symbol size, message size, and IP packet size that provide the best trade-off between coding efficiency and delay/complexity over a range of channel parameters. Then, after setting the message/packet/symbol size the UEP optimization can be done to find the optimal error protection strategies [91].

2.5.1 Symbol Size Selection

The simulations are performed using DVB-H TS packet error traces obtained by realistic DVB-H link layer simulator [53] to determine adequate message/symbol/packet size for systematic Raptor codes (RTC) and RLC. Different RTC and RLC coding configurations are compared in terms of reconstruction performance as well as decoding complexity, where a configuration refers to one pair of source symbol and IP packet size used [91].

After AL-FEC coding (with either RTC or RLC), encoded symbols are grouped into sets of $N_s \geq 1$ symbols and packed into a single IP packet. The number of encoded symbols N_s carried by an IP packet depends on the symbol size l_s, the IP packet size l_{IP} and the set of protocol headers within each IP packet, described as follows. Each IP packet contains a four-byte random number generator seed used to recreate the RTC/RLC encoding coefficients for all the encoded symbols N_s contained within the packet. The remaining symbols derive their coefficients based on the first symbol's Random number generator (RNG) seed. Then, another 60 header bytes comprising RTP/UDP/IP headers are added to form an IP packet (this number can be reduced using robust header compression). Therefore, IP packet size is equal to [91]:

$$l_{IP} = l_{OH}^{(IP)} + N_s . l_s, \tag{2.1}$$

where $l_{OH}^{(IP)}$= 64 bytes.

After IP packetization, at lower DVB-H layers, each IP packet is arranged into a separate link-layer MPE section containing MPE header of $l_{OH}^{(MPE)} = 16$ bytes. MPE section is split into TS packets each with a four-byte header followed by 184 bytes of payload. Assuming that each IP packet with the corresponding MPE section overhead fits into exactly $N_{TS} = \frac{(l_{IP}+16)}{184} TS$ packets, the total overhead per IP packet equals [91]:

$$l_{OH} = l_{OH}^{(IP)} + l_{OH}^{(MPE)} + N_{TS}.4 = 80 + \frac{l_{IP}+16}{184}.4 \tag{2.2}$$

Different IP packet sizes l_{IP} are tested ranging from 300 bytes to over 1450 bytes (the IP packet size limited to 1500 bytes). First, the IP packet size l_{IP} is picked so that, jointly with MPE overhead, it fits a whole number of TS packets, i.e., $l_{IP} = N_{TS}.184 - 16$, where N_{TS} is a positive integer. Then, the available capacity of IP packet is divided into a selected number of N_S equal-length encoded symbols, where the symbol size is determined as $l_s = \left(l_{IP} - l_{OH}^{(IP)}\right)/N_s$ [91].

Table 2-1: System Configurations [91].

FEC	l_{IP} [bytes]	N_{TS}	l_s [bytes]	N_S	$\frac{(l_{OH}+16)}{N_{TS}.188}$ [%]
RTC656	352	2	144	2	23.40
RTC721	720	4	131	5	12.76
RTC1475	1088	6	62	16	9.22
RTC1086	1456	8	87	16	7.45
RTC136	1456	8	696	2	7.45
RLC144	720	4	656	1	12.76
RLC185	1088	6	512	2	9.22
RLC68	1456	8	1392	1	7.45
RLC136	1456	8	696	2	7.45

Let the total source bit budget be *B* bytes; for example, this is the size of a GOP. Then, the number of source symbols in the source message is $k = ceil(B/l_s)$. By varying l_s and l_{IP}, different coding configurations are obtained. Five different RTCs and four RLCs shown in Table 2-1 are tested. Four configurations use l_{IP} in the range of 1400-1500 bytes, two configurations use the IP size of roughly 1000 bytes, while three use very low IP packet sizes. This way low, medium and large IP packet sizes are covered [91]. The postfix number in the FEC scheme's name denotes the source message length *k*.

The last column of the table describes the transmission efficiency in terms of percentage of overhead data within the total transmitted data. From (2.2), it is noted that with increasing N_{TS}, the overhead grows much slower than the total amount of transmitted data, which makes the overall transmission more efficient for larger IP packets [91].

The symbol sizes l_s are generally selected to be small for RTC, so as to yield larger source message size (in conventional FEC, the increase of message length generally improves performance). However, for a fair comparison, RTC with smaller source message sizes is included. The reverse is true for RLC, i.e., larger symbol sizes l_s are selected so as to yield shorter source message sizes k to make the decoding complexity feasible. The selection of different l_{IP} for RTC and RLC is motivated by the need to design and test schemes with different N_{TS}, so that the scheme's response to the TS error trace files could be evaluated over different possibilities. l_{IP} is always picked so that after adding a 16-byte

MPE header, an integer multiple of 184 (payload of a TS packet) bytes is obtained [91].

Let B = 94,400 bytes. Then, for example, in the RLC68 configuration, the source message length of k = 68 symbols results in the symbol size l = 1392 bytes, which is obtained after adding padding zeros to $\lceil 94{,}400/68 \rceil$ = 1389 bytes. In this case N_s = 1 encoded symbol is placed in the IP packet. This symbol gets added with 64 bytes of header data to yield an IP packet of size l_{IP} = 1456 bytes. During transmission, this IP packet occupies N_{TS} = 8 TS packets after having added the MPE header to the IP packet and includes the total overhead of l_{OH} bytes, resulting in the ratio of overhead data in the data stream equal $\frac{l_{OH}}{N_{TS}.188} = 7.45[\%]$ [91].

2.5.2 Simulated EEP Performance of Raptor and RLC

The video sequence *Foreman* in CIF format encoded using the H.264/AVC software JM version 16.2 [54] is used without any error resilient option. All simulations have been performed using a GOP size of 64 frames, one slice per frame, and a frame rate of 25 frames per second. This configuration implies a GOP length of 2.56 seconds of video data and is also used in [52]) [91].

The trace files with the losses of TS packets for different signal-to-noise ratio (SNR) in the DVB-H channel are obtained by realistic link-level simulator, which is confirmed to mimic actual reception in the field [53]. The link-level simulator takes as input TS packet error traces, analyzes them, and generates new TS packet error traces of arbitrary length with similar characteristics. In this sense, the link-level simulator provides a flexible and easy-to-use tool for generating realistic traces, while reducing physical layer simulation times. For generating the input to the link-level simulator, the physical layer simulation model assumes a TU-6 DVB-H channel model with a constant Doppler frequency of 10 Hz, 8k mode, and guard interval 1/4. The simulated modulation is 16-QAM encoded with a convolutional code of rate 1/2. The MPE-FEC error correction is not used [91].

The service bit rate in these simulations was set to 452 kbps. To take into account the varying nature of mobile reception, the video bitrate was limited to 295 kbps, leaving 1/3 of the available bandwidth for transmitting redundant data. The video bitrate also implies a GOP size, and source bit budget B of 94,400 bytes [91].

The results over a range of acceptable code lengths for both codes are reported. Indeed, for RTC configurations with message lengths between k = 136 symbols and k = 1475 symbols are tested. For RLC, on the other hand, configurations between k = 68 symbols and k = 185 symbols are tested, since a higher k would result in unacceptable decoding complexity (see [20]).

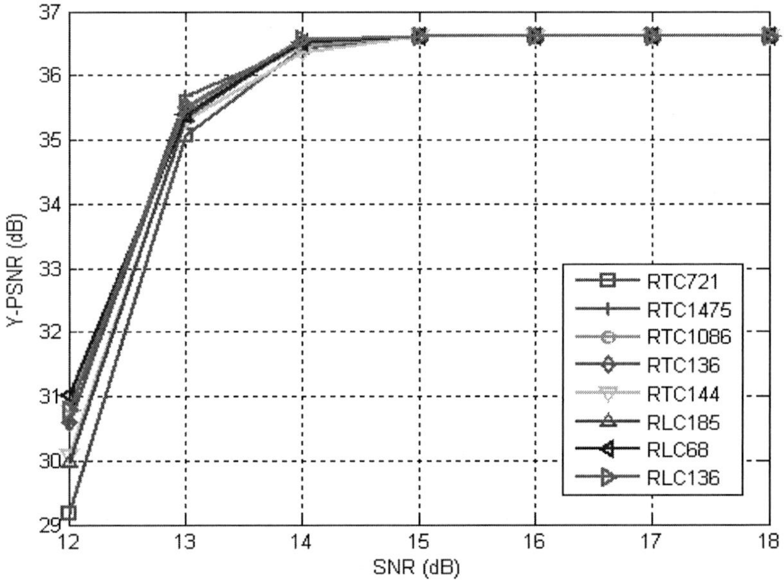

Figure 2-4: PSNR comparison between four selected EEP configurations: Y-PSNR vs. SNR in the DVB-H channel [91].

The expected peak signal-to-noise ratio (PSNR) averaged over all frames in the sequence as a criterion for quality comparison is used. To compare the quality for each of the used configurations, the overall data rate was fixed to be 452 kbps and each configuration is subjected to the same loss patterns. [91], [55] © 2011 IEEE. The results are shown in Figure 2-4.

It can be seen that several configurations, RTC1475, RTC1086, RTC136, RLC68, and RLC136 provide the best PSNR performance. The worst performing sequence is RTC656 (not shown in the figure) and then RTC721 which have the heaviest data overhead. In particular, for SNR = 12 dB, RLC68 is the best performing with PSNR = 31.02 dB while the best performing RTC, RTC1086, is lagging behind for 0.16dB. At SNR = 13 dB, RTC1475 is the best performing with PSNR = 35.69 dB while RLC68 performs for only 0.09 dB worse, and this gap is similar at SNR = 14 dB. At SNR = 15 dB all eight schemes reach error free PSNR performance of 36.61 dB [91].

The two best performing schemes are adopted from each code class (RTC and RLC), i.e., RTC136, RTC1086, RLC68, and RLC136, having in mind that shorter message sizes are preferable option if performance is not severely affected [91].

2.6 DP H.264/AVC Video Broadcasting with UEP AL-FEC

In this section, the DPs are organized into classes of different importance. Intuitively, Instantaneous Decoder Refresh (IDR) and DP A is always placed in so called High Priority Class (HPC), because the remaining data cannot be decoded unless IDR and DPs A are correctly received [91].

Furthermore, two schemes are created, RLC-IAB, where IDR, DP B and DP A are placed in the HPC and treat DP C as the less importance class called Low Priority Class (LPC). The size of layer HPC is then roughly 70% of the total size. However, the second layer provides a significant performance improvement of over 7 dB for the Foreman sequence. In the second scheme, RLC-IA, only IDR and DP A are placed into HPC, and DP B and DP C comprise LPC. The size of HPC is then about 55% of the total stream size [91].

In order to demonstrate the effect of removal of DP A and DP B, frame 45 for both RLC-IA and RLC-IAB has been extracted and is shown in Figure 2-5. The advantage of RLC-IA is a smaller base layer (with low quality) with less decoding failures though UEP.

(a) Original sequence, PSNR 35.64 dB.

(b) RLC-IAB, PSNR 28.64 dB.

(c) RLC-IA, PSNR 24.51 dB.

Figure 2-5: Frame 45 of the Foreman sequence for various configurations.

Table 2-2: Four Selected Coding Configurations for the UEP Analysis [91].

Scheme	No of Symbols		Symbol Size [bytes]	No. Symbols in IP Packets
	HPC	LPC		
RTC1086	776	305	87	16
RTC136	97	39	696	2
RLC68	49	19	1394	1
RLC136	97	39	696	2

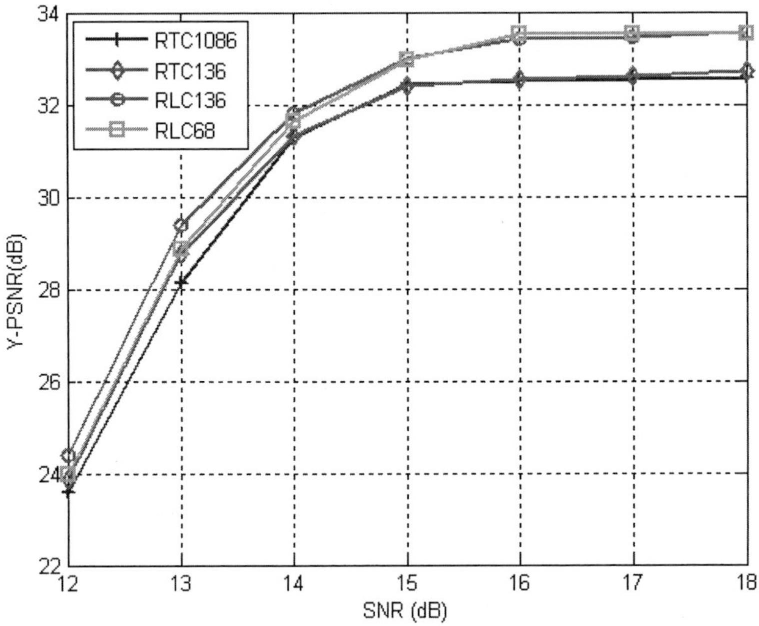

Figure 2-6: PSNR comparison between four selected RTC and RLC configurations: Y-PSNR vs. SNR in the DVB-H channel [91].

Figure 2-6 shows frame-average PSNR vs. channel SNR for the four selected RLC-IAB configurations, shown in Table 2-2, when transmission data rate is fixed at 383 kbps and with protection probability of HPC of 75%. The results with values of SNR that were available from the link-level simulator are presented [91].

For simulating each scheme, the TS packets (as determined by the data rate) are transmitted and are subjected to the loss pattern. Loss of an IP packet is assumed if any of its enclosed TS packets is lost.

Although the difference in performance is not very significant, however, RLC136 is the best performing configuration for all SNRs. The PSNR difference between RLC68 and RTC136 is marginal. However, RTC1086 provides the worst overall performance [91]. Similar results were obtained for other UEP schemes (see [58]).

Having in mind the limited complexity of GE decoding for RLC68 (due to small code length), very small delay introduced, and competitive performance results, it is concluded that it is a good candidate for real-world implementations.

Thus, all further simulations are presented for different UEP configurations obtained with RLC68 [91].

If channel characteristics are known, it is possible to provide optimal allocation of redundancy to different source priority classes. For this task, the expected PSNR is maximized at channel packet loss rate ϵ using analytically computed probabilities of decoding error performance.

That is,

$$max_{\pi} PSNR(\pi, \epsilon) = \sum_{i=0}^{L} P(i, \epsilon) psnr(i) \tag{2.3}$$

where L is the number of classes, $P(i, \epsilon)$ is probability that class i will be the highest class recovered at channel packet loss rate ϵ, $P(0)$ is the probability that nothing is recovered, $psnr(i)$ is PSNR of the reconstruction if all classes up to and including class i are recovered ($i = 1,...,L$), π is an L-tuple vector of window selection probabilities that determines the UEP allocation scheme, and $PSNR(\pi, \epsilon)$ is the expected PSNR when UEP scheme π is used [91].

Analytical expressions for probabilities $P(i, \epsilon)$ assuming a random channel loss model, for both NOW and EW schemes [14] can be found in [50].

Above, however, it is assumed that channel characteristics are known at the transmitter, which is unrealistic assumption in DVB-H. Thus, instead, the average performance over a range of expected packet loss rates is maximized [91]:

$$\underset{\pi}{max} PSNR(\pi) = \sum_{\epsilon_j \in [\mathrm{I}]} \sum_{i=0}^{L} P(i, \epsilon_j) psnr(i), \tag{2.4}$$

where [I] is the set of probable packet loss rates. This way, it is aimed to find a scheme that performs in average the best over a range of channel conditions. Since the number of possible UEP schemes is finite and small

(L = 2 or 3), in the following we maximize (2.4) using exhaustive search, which is very fast [91].

2.6.1 Performance Evaluation of UEP NOW and EW RLC

Next, for RLC68, NOW and EW schemes are compared. We used four different video sequences: the standard CIF Foreman and Coastguard, QCIF Bus and 4CIF City sequence. The same source configuration as in the previous section is used. For illustration, the employed EW schemes for the Foreman sequence are shown in Table 2-3. The symbol size for all the schemes is 1394 bytes. EW1 refers to the single layer EEP scheme (with all parameters set as in the EW2 scheme) [91].

Table 2-3: Configurations and the Window Structure for RLC68 [91].

Scheme	Window	No. Of Symbols	No. Of IP Packets	No. Of TS Packets
EW1	W1	68	68	544
EW2(RLC-IAB)	W1	49	49	392
	W2	19	19	152
EW2(RLC-IA)	W1	37	37	296
	W2	31	31	248
EW3	W1	37	37	296
	W2	12	12	96
	W3	19	19	152

Results for the Foreman sequence for four NOW UEP schemes, four EW2 UEP schemes and the EEP scheme of the RLC68-IAB scheme at data rate of 383 kbps are shown in Figure 2-7. UEP_xN, x = 60, 75, 85, or 100 refers to a NOW UEP scheme with x % probability of selection for HPC. Similarly, UEP_xE, is an EW scheme with x % probability of selection for HPC. We show frame-averaged PSNR vs. SNR in the DVB-H channel [91].

As can be seen from the figure, the EW UEP schemes outperform EEP and NOW schemes for all SNRs. The EEP scheme outperforms the best NOW scheme for high SNRs, while for low SNR, the performance of the NOW UEP schemes is better [91].

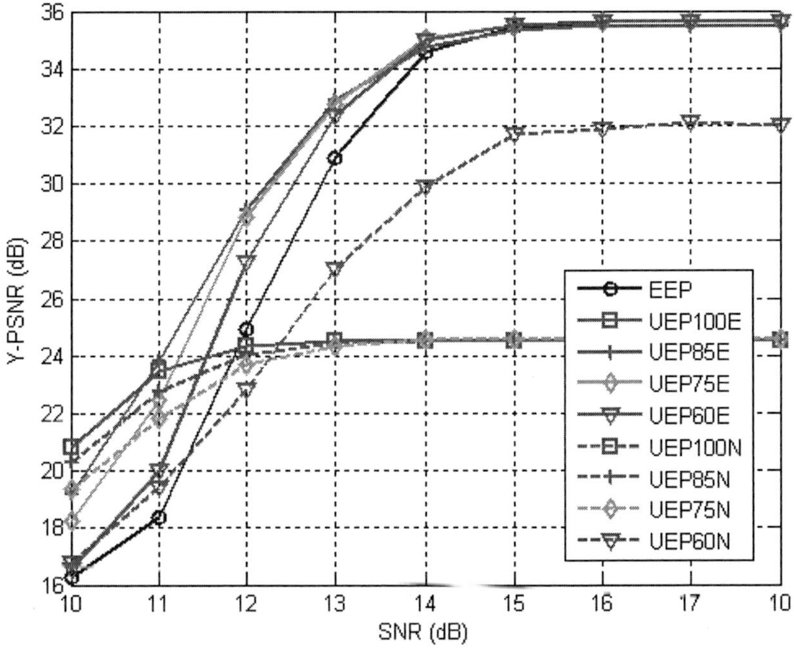

Figure 2-7: PSNR vs. SNR in the DVB-H channel for four different NOW and EW UEP schemes and the single-layer EEP scheme of RLC68-IAB at data rate of 383 kbps for Foreman sequence [91].

Superiority of the EW schemes over NOW schemes is clear. The main reason for performance gains of EW compared to NOW lies in the fact that NOW attempts to select symbols from either class in a mutual exclusive manner, in the sense, that if the protection for HPC is increased, this directly detriments LPC. In case of the EW schemes the windows are nested and correctly received packets from smaller windows may be used in recovery of larger windows. The optimal EW scheme in terms of maximizing (2.4) was UEP(85,15), whereas the optimal NOW scheme was UEP(60,40). In both cases it is assumed that SNR values are integers between 10 dB and 18 dB, average packet loss rates for each SNR is determined, and maximization is performed as in (2.4) [91].

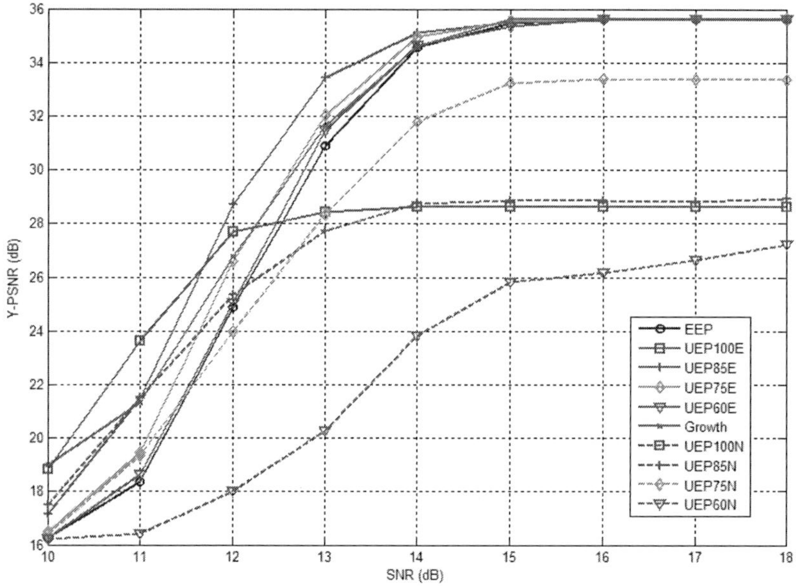

Figure 2-8: PSNR vs. SNR in the DVB-H channel for four different NOW and EW UEP schemes and the single-layer EEP scheme of RLC-IA at data rate of 383 kbps for Foreman sequence. 'Growth' denotes the scheme of [47]. [91].

The PSNR results obtained by varying the SNR over its entire available range and keeping the data rate fixed at 383 kbps are shown in Figure 2-8 for RLC-IA. Similar conclusions as for the RLC-IAB scheme can be taken. NOW schemes again lag behind EW for higher SNRs [91].

It can be seen from the two figures that the optimal EW2 UEP(85,15) scheme is superior to all the other schemes including EEP, except at low SNRs, when UEP(100,0) is better for only up to 1dB. This justifies the proposed optimization strategy and shows robustness of the EW RLC UEP schemes over large range of SNRs. It can also be seen that the proposed optimized UEP schemes outperforms the scheme of [47] for the entire range of SNRs. The scheme of [47] uses a nested window structure to realize UEP where the inner window contains IDR, DP A and DP B, while the outer window contains also DP C. The protection is equally divided between the two windows. Since Growth codes, used in [47], have similar performance to Raptor codes, and hence RLC codes, the advantage of the proposed solution comes from the optimal parameter selection (Figure 2-8) [91].

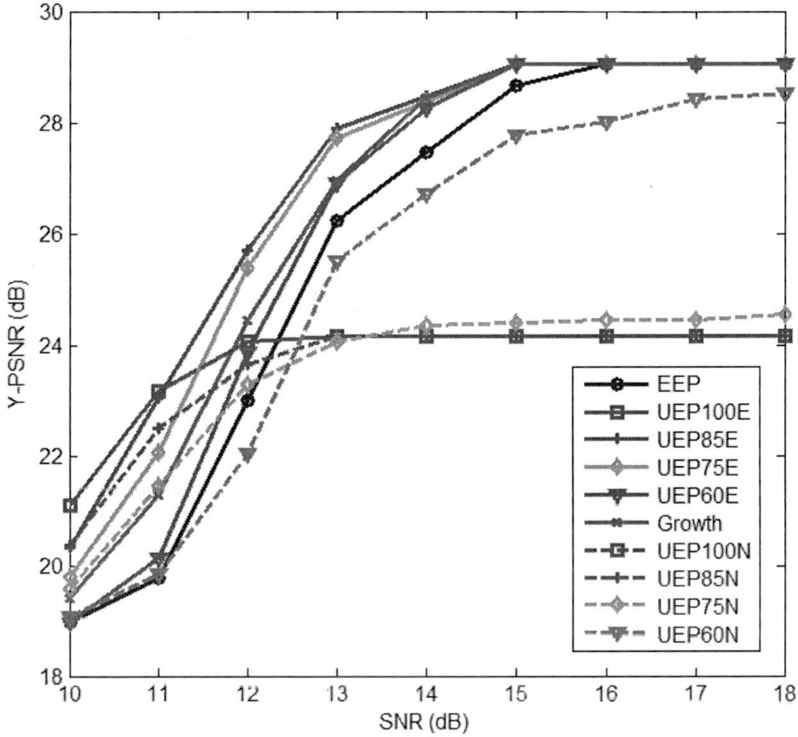

Figure 2-9: PSNR vs. SNR in the DVB-H channel for four different NOW and EW UEP schemes and the single-layer EEP scheme of RLC-IA at data rate of 383 kbps. The standard CIF Coastguard video sequence is used. 'Growth' denotes the scheme of [47]. [91].

Similar behavior can be observed from the next figure that shows results for the CIF Coastguard sequence. UEP85E is the optimal scheme for SNR higher than 11dB. The scheme of [47] loses roughly 1-2dB compared to the UEP85E at low SNRs. Figures 2-10 and 2-11 show results obtained with the QCIF Bus and 4CIF City sequence, respectively. It can be seen from the figures that the optimized EW UEP scheme is the most efficient even for different video resolutions [91].

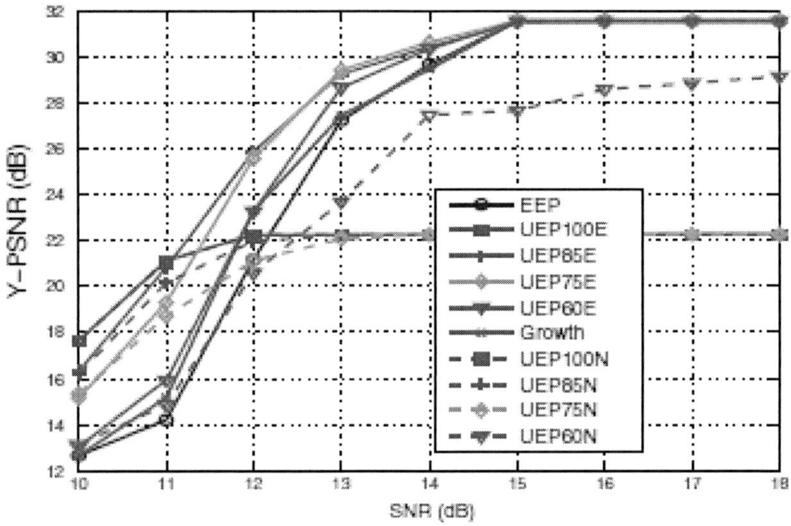

Figure 2-10: PSNR vs. SNR in the DVB-H channel for four different NOW and EW UEP schemes and the single layer EEP scheme at data rate of 383 kbps. The standard QCIF Bus video sequence is used. 'Growth' denotes the scheme of [47]. [91].

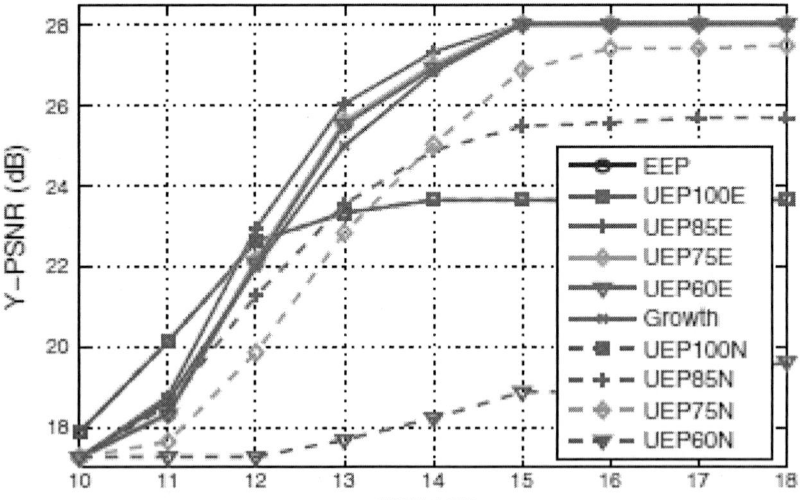

Figure 2-11: PSNR vs. SNR in the DVB-H channel for four different NOW and EW UEP schemes and the single layer EEP scheme at data rate of 383 kbps. The standard 4CIF City video sequence is used. 'Growth' denotes the scheme of [47]. [91].

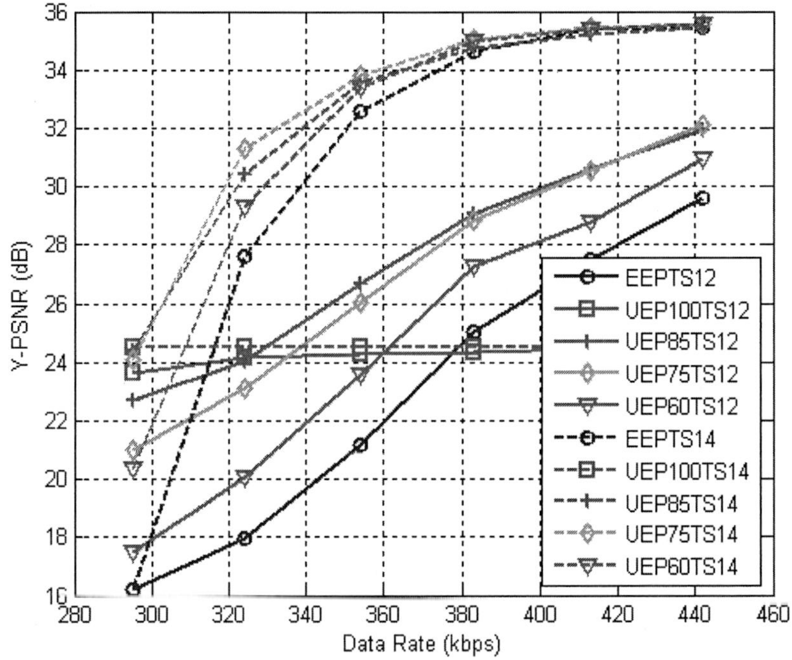

Figure 2-12: PSNR vs. data rate for EW2 (RLC-IA) and EW1 (single-layer) at SNR 12dB and 14dB for Foreman sequence [91].

Figure 2-12 shows frame averaged PSNR vs. transmission data rate for EW1 and EW2 (RLC-IAB) at two different SNRs for the Foreman sequence. Single-layer EEP refers to EW1 scheme in Table 2-3, whereas EEPTS12 and EEPTS14 refer to an EEP scheme at SNR=12 and 14 dB, respectively. Similarly, UEP60TS14 represents a UEP scheme with 60% selection probability for HPC at SNR = 14 dB. The results for UEP75TS12/14 = UEP(75, 25) and UEP(85, 15) provide significant improvement over the single-layer case. As can be seen from the figure, UEP(85, 15) provides better results overall, whereas UEP(100, 0), in which only W1 is protected, provides good results only at low data rates. This is expected, because in this case, there is not enough bandwidth to transmit both classes, thus, it is better not to waste the bandwidth in trying to recover the LPC and use all the resources to ensure recovery of the HPC. Note also that UEP(100,0) reaches almost saturation for higher rates since then, HPC is always delivered and the remaining resources are not used [91].

Similar results were obtained for Coastguard sequence and are shown in Figure 2-13. Except for the fact that the overall PSNR for Coastguard sequence is lower as compared to the Foreman sequence, the general performance of the used configurations is the same. This confirms the analysis and results as given for Figure 2-12 for Foreman sequence.

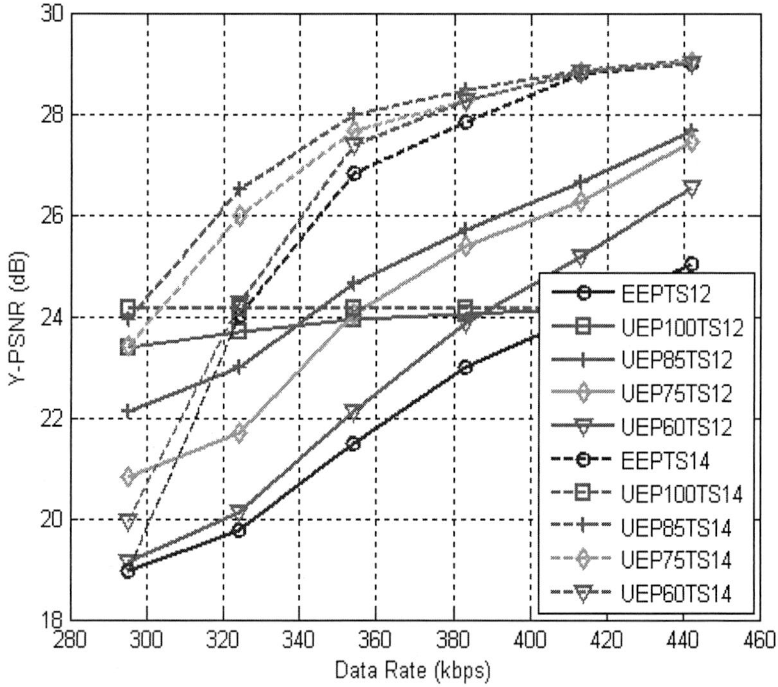

Figure 2-13: PSNR vs. data rate for EW2 (RLC-IA) and EW1 (single-layer) at SNR 12dB and 14 dB for Coastguard sequence.

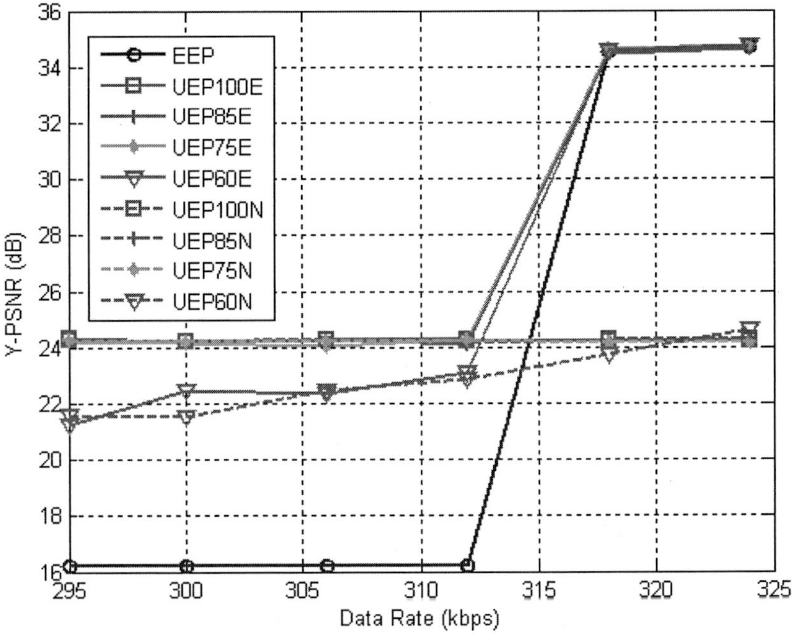

Figure 2-14: PSNR comparison between different RTC and RLC configurations: Y-PSNR vs. Data rate in the DVB-T2 channel for Foreman sequence [91].

2.6.2 Performance Evaluation of EW-RLC Based Solution with DVB-T2

DVB-T2 Lite profile is intended to allow simpler receiver implementations for very low capacity applications such as mobile broadcasting, although it may also be received by conventional stationary receivers. The SNRs for the simulations have been obtained through DVB-T field measurements performed in Turku in August 2011, where each SNR value was measured with 250 ms intervals (corresponding to the maximal DVB-T2 frame duration). The length of the SNR series corresponds to 1040.5 s of reception of a DVB-T2 signal. The SNRs have been used for running DVB-T2 simulations with the following DVB-T2 Lite configurations: 16-QAM and 64-QAM modulations, 16200 bits LDPC codewords with rate 1/2, 8k FFT size, 1/4 guard interval, and TU-6 channel model with perfect channel estimation. By using the SNR's from the DVB-T field measurements as input to DVB-T2 simulations, realistic time variant behaviour of mobile reception was obtained [91].

With these settings, each time interleaver (TI) frame contained 64 or 97 FEC frames for the 16-QAM and 64-QAM simulation scenarios, respectively. The FEC frames have been decoded and a baseband frame error trace has been collected. The baseband frame error traces have then been converted to TS packet error traces, taking into account that the size of each baseband frame is 7032 bits [91].

Figure 2-14 shows the frame-averaged PSNR vs. data rate for EW1 and EW2 (RLC-IA) for 64-QAM. In case of NOW-RLC except for UEP60N, all schemes suffer an overprotection of HPC which is detrimental to overall performance. Hence, these schemes get constrained to a PSNR performance of about 24.51 dB, the best achievable with HPC alone. For EW-RLC schemes, the performance is poor at 295 kbps and thereafter all the schemes perform equally well at the increased data rate [91]. Similar results were obtained for the Coastguard sequence and are shown in Figure 2-15.

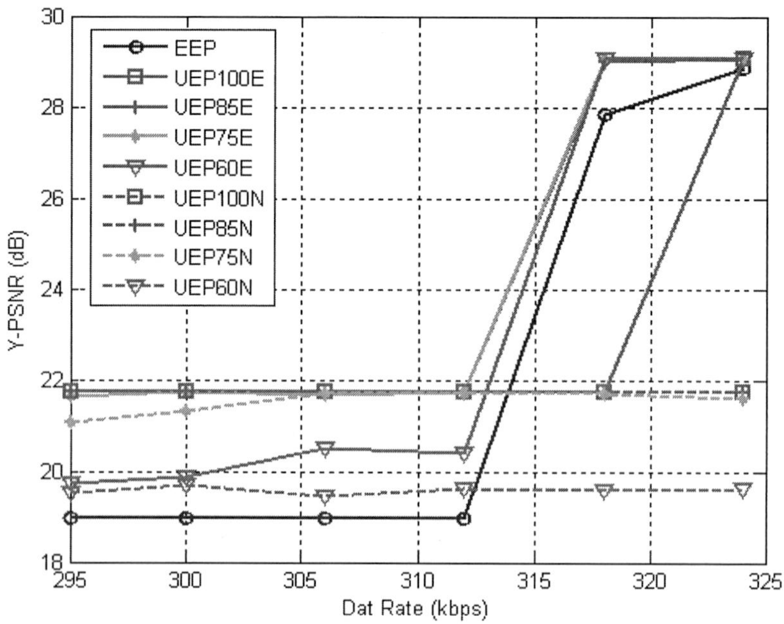

Figure 2-15: PSNR comparison between different RTC and RLC configurations: Y-PSNR vs. Data rate in the DVB-T2 channel for Coastguard sequence.

2.7 Slice Partitioned H.264/AVC Video Broadcasting with UEP RLC

In this section, the video data is partitioned based on the slicing feature of H.264/AVC in order to provide UEP with EW-RLC. The process of packetization etc is same as that for DP.

2.7.1 Overall System and Measurement Setup

In order to enable progressive recovery, the video sequence is encoded with the slicing feature with six slices per frame. The MBIU feature is used to additionally limit the effect of error propagation and each slice contains a fixed number of MBs. After the H.264/AVC encoding, the video data in which each frame including the Instantaneous decoder refresh (IDR) is divided into six slices is obtained. The priority of each slice is obtained by dropping it from the GOP data and measuring the resulting PSNR by actual decoding [68] © 2011 IEEE.

The slice sorting can also be done as an analytic process as described in [59], which explains the slice sorting by relevance (SSR) algorithm and how it can be used to prioritize each slice. After the slice sorting in [59], different priority slice partitions are protected with RS codes [68] © 2011 IEEE.

The loss of a slice has an error propagation effect which adds to the GOP distortion, but if the previous frame is similar (low motion), such effect is minimal. This fact is used to create a prioritized and segmented video data. The cumulative PSNR of the GOP is measured with one slice dropped out, starting at the first P frame. The GOP structure is always IPPPP… After having obtained the cumulative PSNR values for each slice, the deviation from the full decoding PSNR of the GOP is measured as shown in Figure 2-16 for the first GOP (having 64 frames) of the standard CIF Foreman sequence. As can be seen from Figure 2-16, some slices have very limited contribution. It is such slices which can be assigned less protection in order to afford a higher degree of protection for the more important slices. On the other hand, loss of one slice causes a PSNR drop of 3.5 dB [68] © 2011 IEEE.

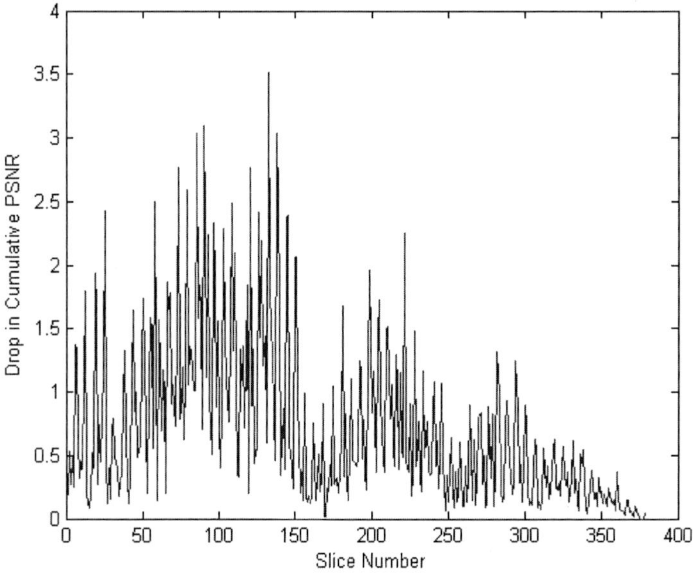

Figure 2-16: Drop in Cumulative PSNR vs a missing slice for the first GOP of Foreman sequence. The PSNR without missing slices is 35.63 dB. [68] © 2011 IEEE.

The scheme that is proposed here is to choose a threshold T based on the PSNR contribution. All slices, whose contribution is above the threshold, get assigned to the higher priority segment/layer. Similarly, creation of N segments requires N-1 thresholds. Additionally, any frame which does not have any slice selected in the high priority layer based on the threshold alone will have at least one representative slice included to offset the effects of error propagation and to improve PSNR. Such frames occur mostly towards the end of GOP as can be seen from Figure 2-16. Note that the process of slice prioritization is of low complexity since the encoder already has all the information used for rate-distortion optimization [68] © 2011 IEEE.

In the proposed EW2 RLC scheme, with a single threshold T, two windows, W1 and W2 are created. In the proposed EW3 scheme, two threshold values are utilized to create three windows, W1, W2, and W3. Note that for both these schemes, all slices of IDR frame are placed in W1 being the most important data for video reconstruction [68] © 2011 IEEE.

The data in each window is divided into equal-length source symbols. The length of the source symbols determines the number of source symbols contained in the source message, k. The symbol size has been set to 1394 bytes because large source symbols are preferred as they yield small values of k which is necessary to keep the RLC decoding complexity acceptable. RLC configuration with 1394 symbol size achieves the best trade-off between performance and decoding complexity over a range of transmission rates [68] © 2011 IEEE.

After AL-FEC coding, one encoded symbol (together with RTP/UDP/IP headers) is placed in an IP packet. The IP packets are then placed into the MPE frames where each IP packet is encapsulated within a single MPE section with its own overhead containing the error-detection field. The MPE frame is transmitted within a single DVB-H physical layer transmission burst [68] © 2011 IEEE.

Table 2-4: Relative Partition Sizes for the First GOP of the Foreman Sequence. [68] © 2011 IEEE.

<table>
<tr><th>Partition</th><th>Size (bytes)</th><th>Size (%)</th><th>Cumulative PSNR [dB]</th><th>EW3</th><th>EW2</th></tr>
<tr><td>SP1</td><td>52,980</td><td>55.65</td><td>22.09</td><td>W1</td><td rowspan="2">W1</td></tr>
<tr><td>SP2</td><td>15,320</td><td>16.16</td><td>26.84</td><td>W2</td></tr>
<tr><td>SP3</td><td>26,480</td><td>28.20</td><td>35.63</td><td>W3</td><td>W2</td></tr>
<tr><td>Total</td><td>94,780</td><td>100</td><td>35.63</td><td colspan="2">94,780 bytes</td></tr>
</table>

2.7.2 Results and Analysis

In all the simulations the video sequence *Foreman* in the CIF format is encoded using the H.264/AVC software JM version 16.2 [54]. All the simulations are performed using a GOP size of 64 frames, with frame structure IPPPP…, six slices per frame with 66 MBs per slice, and a frame rate of 25 frames per second (fps). This configuration implies that each GOP has an average size of 94,400 bytes corresponding to 2.56 sec of video data, and thus the source data rate is roughly 295 kbps. This ensures that sufficient amount of video data is transmitted in one burst of the DVB-H transmission. The partitions with their relative sizes and the PSNR contribution are shown in Table 2-4. The partitions SP1 to SP3 are obtained using threshold values on the PSNR difference from a full decoding (without any losses) PSNR [68] © 2011 IEEE.

Table 2-5: Configurations and the Window Structure. [68] © 2011 IEEE.

Configuration	Sym. Size (bytes)	Window	No. of Symbols	No. of IP Packets	No. of TS Packets
EW2	1394	W1	49	49	392
		W2	19	19	152
EW3	1394	W1	38	38	304
		W2	11	11	88
		W3	19	19	152

Two windows (EW2) and three windows (EW3) are created by grouping the partitions as shown in the table. The GOF video data is divided into the source blocks containing 68 symbols, which are then assigned to various windows as shown in Table 2-5. The symbol size used and the number of symbols in each window for the different schemes are also shown [68] © 2011 IEEE.

The configurations shown in Table 2-5 are simulated with 1000 runs each for DVB-H channel SNR equal to 12 dB. To model the channel accurately, TS packet losses are represented based on trace files for different channel SNRs for the DVB-H channel with real-world measurements conducted at [52]. The simulations have been performed for different data rates to evaluate the EEP and UEP performance of EW2. The base data rate is 295 kbps, and the successively higher data rates are obtained by successively adding 10% additional symbols. The highest rate obtained is 442 kbps, which is 1.5 times higher than the base rate [68] © 2011 IEEE.

In case when even the first window W1 does not get decoded, the entire GOP is considered to be lost. The PSNR for such cases is obtained by using the last frame of the previously decoded GOP to replace all frames of the lost GOP. The various configurations are used to create different UEP schemes based on protecting the constituent windows with different protection, based on probabilistically selecting a window for each output symbol at the transmitter. An increase in the selection probability of W1 will improve its robustness at the cost of a decrease in robustness of the succeeding layer(s) [90] © 2011 IEEE. The EEP scheme is the case where only the largest window is selected with 100% probability. This means that all of the data is protected with no preference for the data considered important, i.e., window W1 [68] © 2011 IEEE.

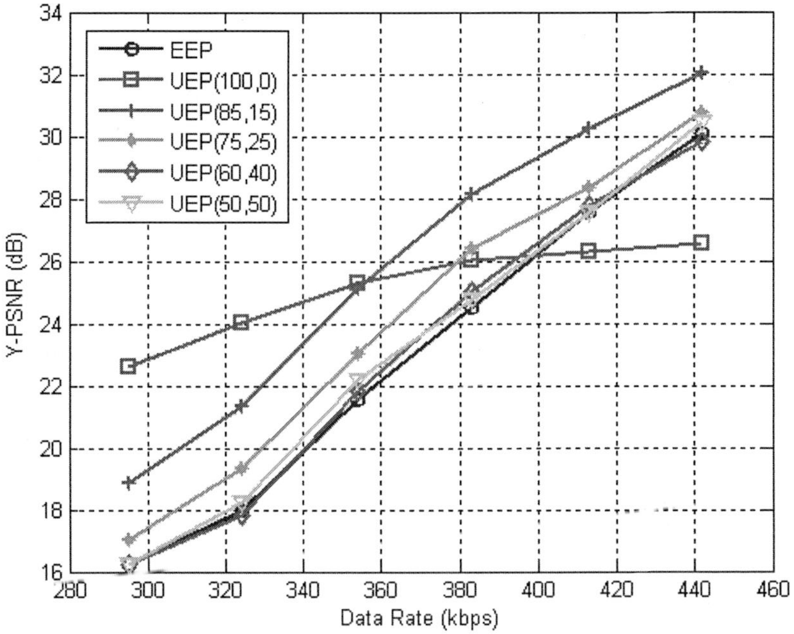

Figure 2-17: PSNR for EW2 at SNR 12 dB [68] © 2011 IEEE.

Figure 2-17 shows PSNR vs. data rate for EW2. The numbers shown in brackets represent the selection probability of each of the two windows, e.g., UEP (85, 15) represents a code in which a symbol from W1 will be selected for transmission with probability 0.85 [68] © 2011 IEEE.

The performance of UEP (75, 25) and UEP (85, 15) is much better as compared to the EEP. As can be seen from the figure, UEP (85, 15) provides better results overall for higher data rates. UEP (100, 0) is a scheme in which only W1 is protected. The scheme is constrained in that it cannot achieve any better than 26.84 dB (see Table 2-4). The results are thus better than other schemes only at low data rates. However, the decoding failures, i.e., when the GOP data fails to be decoded, are much less for UEP (100, 0), which is good as this ensures that for each GOP, data is received, though at basic quality level [68] © 2011 IEEE.

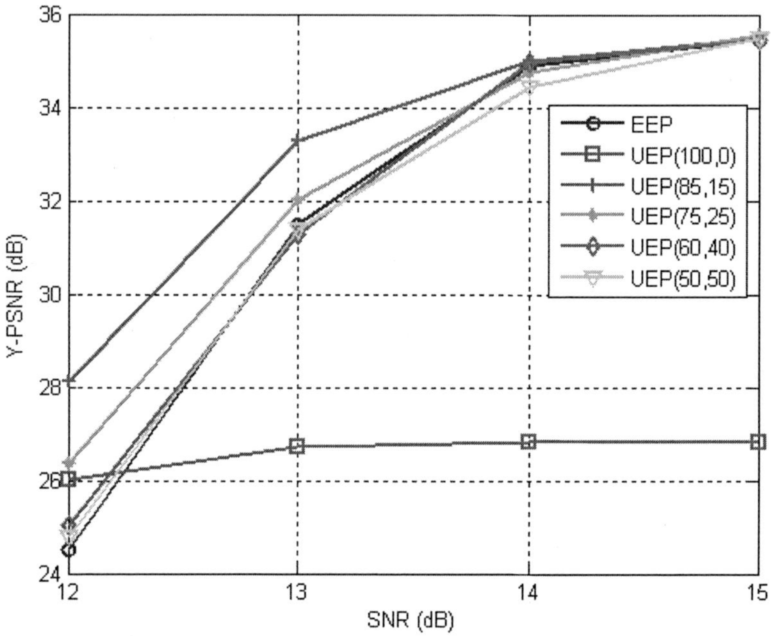

Figure 2-18: PSNR at different SNRs for 383 kbps [68] © 2011 IEEE.

Due to small code length for RLC68, the decoding complexity is minimal, which makes it a good candidate for real-world implementations such as the one described in [20]. The results of Figure 2-17 are similar to those in earlier sections where DP feature of H.264 encoder is used to create the windows [68] © 2011 IEEE.

The PSNR for EW2 at 383 kbps for various SNR values is shown in Figure 2-18. At SNR 12 dB, some schemes perform worse than UEP(100,0) which is only protecting W1, which signifies that with such low probabilities of selection, i.e., 50 and 60%, even W1 is not being protected adequately. On the other hand, UEP(85,15) has better performance even at such low SNR. All schemes reach highest PSNR of 35.63 dB at SNR 15dB.

The degree of protection provided with selection probability in UEP schemes can thus be made dynamic to change on GOP basis [68] © 2011 IEEE.

2.8 Conclusion

Comparative performance results for Raptor and RLC codes have been provided at various data rates. The performance of EW-RLC schemes is significantly better than the corresponding NOW-RLC schemes for H.264/AVC partitoned video with DP feature.

The results and analysis are also substantiated by using another of error resilience feature, i.e., slicing. By choosing similar configurations as that of DP the results with slicing are provided which confirm the earlier results. For the slice-partitioned schemes the performance of RLC codes using EW-RLC has been analyzed and compared to the case of single-layer coding. EW2 performs up to 5 dB better compared to the single-layer case and shows robustness to channel fluctuations.

In the next chapter, the DP and slicing schemes are used to propose a rate-adaptive solution for reliable video transmission.

CHAPTER THREE

RATE ADAPTATION TECHNIQUES FOR VIDEO STREAMING

3.1 Introduction

In this chapter, a reliable video communication system is proposed which uses DP and slicing feature of H264/AVC to create a layered stream, and LT codes [8] for erasure protection. The proposed scheme is an adaptive low-complexity solution to varying channel conditions appropriate for unicast scenarios. The comparison of the results of the proposed scheme is also provided for slice-partitioned H.264/AVC data. Simulation results show competitiveness of the proposed scheme compared to optimized unequal and equal error protection solutions. The simulation results also demonstrate that a high visual quality video transmission can be maintained despite the adverse effect of varying channel conditions and the number of decoding failures can be reduced.

Fountain codes [7], [9] are rateless and in non time-constraint applications can generate as many encoded packets as needed. The amount of additional packets transmitted is the redundancy which is necessary for decoding to succeed and can be adjusted to combat different channel conditions. In bandwidth-limited wireless networks it is important to keep the introduced redundancy to a minimum. Thus instead of targeting the worst possible channel conditions, the redundancy should be adaptively adjusted according to the varying channel conditions via dynamic source-channel coding. LT codes are used in this chapter due to their design and implementation simplicity. However, LT codes have a higher computational complexity, $O(k \log_e k)$ (k is the message length) than Raptor codes, O(k) [92].

In this chapter three schemes are evaluated, namely fixed-source rate EEP, fixed source-rate UEP, and source rate-adaptive schemes in the context of H.264/AVC data prioritization. Different error resilience options are used to generate encoded segments that can be dropped to adjust the rate or provide UEP based on their importance. Error-resilience and concealment features are utilized, which are designed to make the video less vulnerable to the effects of lost data, and then the video is compressed at a higher source rate allowing for some decoding errors [92].

DP [2] is a low cost error resilience feature, supported by the extended AVC profile, which can be exploited to introduce a layered structure in H.264/AVC. Besides DP, it is possible to partition a frame into a fixed number of *slices*, which are of different importance to the video reconstruction. Thus, similar to DPs, the slices can be aggregated into different priority classes, with the higher priority classes containing slices that have higher contribution to the reconstruction. Such prioritization can make the sliced video data amenable to UEP and rate adaptation. DP has low overhead as its structure is determined in advance, whereas slicing generally requires a slice group map [92].

In this study the data in various partitions, based on DP or slicing, is treated as segments. Then, based on the available channel conditions, namely, the available bandwidth and the estimated wireless channel loss rate, the data from each of the segments is selected for FEC coding. The optimized EEP and UEP are compared with the source-adaptive scheme that is termed rate adaptive selective segment assignment (RASSA), and pros and cons of them are identified in channel mismatch scenarios.

A lot of work has been done on joint source-channel coding; see [61] for a review. In the domain of rateless source-channel coding, in [21], a class of UEP codes, called Expanding Window Fountain (EWF) codes is used for UEP of scalable video. The UEP with LT codes was first proposed in [65]. In [62], unequal protection has been proposed for video communications by duplicating the information symbols. In [63], unequal Growth codes have been proposed in which the degree for each new encoding symbol increases corresponding to the number of packets the receiver has already received. The application of PET to UEP video data is provided in [66]. An adaptive rateless coding for DP AVC coded video has been proposed in [64]. The system uses intra-coded macroblocks (MBs) in each frame; some additional redundant data is piggybacked onto the on-going packet stream [92].

Priority encoded transmission (PET) [60] is another FEC scheme which can be used to provide UEP for video data. The message gets encoded into packets depending on the laid down priorities and thus provides an effective solution to offset errors. In [67] channel coding is combined with additional duplicate packets.

The proposed method differs from the earlier work in that it increases the protection for the transmitted data by not selecting whole of the data for transmission. A scheme has been proposed to decode video even when the rateless decoding fails using packetization information. This is made possible by passing a video-table to the decoder containing the DP type and size information. Thus, the DPs with all or part of their data missing are discarded before the H.264 decoder tries to decode the data. It is important to note that without such information the decoding will fail on encountering such missing data.

The segmentation of video data facilitates a layered coded video that might be preferable to the H.264 Scalable Video Coding (SVC) extension [4] in some applications, since it complies with the AVC standard, provides scalability, and more robust output to packet losses than SVC. The proposed scheme can be applied in multicast scenarios with heterogeneous receivers, in which case a receiver can terminate reception and decoding of segments after having received data compatible with its processing power and memory [92].

The rest of the Chapter is organized as follows. Section 3.2 describes the proposed system. The proposed rate allocation algorithms are described in Section 3.3. The results and analysis are in Section 3.4. Finally, the conclusion is provided in Section 3.5.

3.2 Proposed System

In this section the proposed system is described that segments encoded video and provides equal or unequal error protection. It is to be noted that an LT packet size of 70 bytes is proposed, thus when an LT packet is lost, the corresponding DP/slice needs to be dropped as well. The DPs/slices which are received and hence decoded may have some intervening DP/slice missing. Although the results are not provided for comparison of the proposed prioritization scheme with other types of channel coding schemes (for example, using different types of frames in a GOP with more protection allocated to more important frames) however, it can be argued that such frame-based segregation into UEP classes will provide poor results as compared to DP/slice based classification. The reason is that with such frame based classification, loss of a single LT packet from within a frame will potentially make it useless. The probability of an LT packet being missing from a frame is more (at least three times) as compared to DP/slice based partitions. Moreover, DP/slice partitioning being error-resilience schemes, they also mitigate the error propagation to subsequent frames, whereas comparably a lost frame will have more drastic degrading effect.

First, a system is described that forms layered outputs using the DP feature. Then, the system that exploits slicing instead of DP is presented.

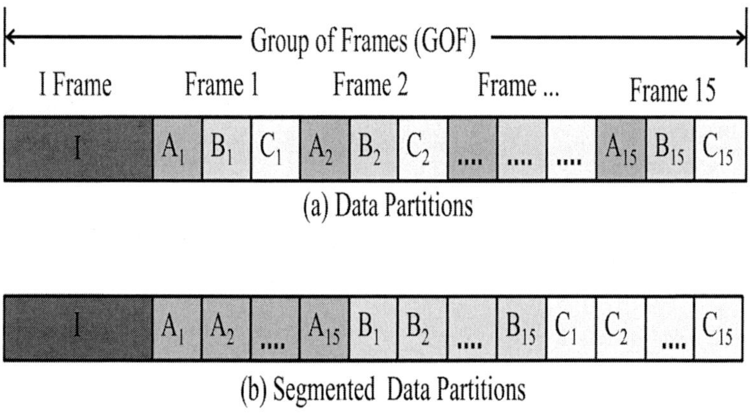

Figure 3-1: (a) Data Partitions. (b) Segmented Data Partitions.

3.2.1 Protection of DP-based AVC Video

The IDR frames are used without any partitioning. The video data of each non-IDR frame is divided into three data partitions by the H.264/AVC encoder. This partitioned data needs to be aggregated together to enable UEP. The structure of a segmented video is shown in Figure 3-1. The figure shows the DP A, B, and C together with IDR frame. (Note that the first non-IDR frame is denoted as A_1, B_1, C_1, and so forth.) Next, the partitions are prioritized and all DP As, Bs, and Cs are grouped together to form effectively three segments or layers as shown in Figure 3-1 (b). Note that by receiving only I/ IDR frame and DP A_1-A_n, the decoder will still be able to decode all n frames within the GOP, though at reduced quality. Further segmentation is not restricted to be done at the aggregate partition boundaries only. That is, if all IDR and DP A are sent as the first segment, then any number of DP B and DP C partitions can be selected for transmission in the second segment. This provides flexibility that enables a fine-grained layered structure as a large number of reconstruction rate points become available, which can be matched to the channel statistics with a very fine control over video reconstruction quality. The DP B and DP C by virtue of having been aggregated are already in their priority order for reconstruction. The layer with important data (IDR and DP A etc.) is termed as High priority layer (HPL), whereas the remaining data is placed in Low priority layer (LPL). In the proposed scheme, intra-refresh MBs are not used but instead periodic I frames are assumed [92].

The segmented data partitions are next protected by FEC codes applied on each GOP independently. To achieve UEP, each segment should be protected according to its importance using different amount of redundant symbols. To accomplish that, the FEC encoding process adds an important initial step, that is, to first select a segment from which the encoded symbol is to be generated determined by "selection probability" of a segment which is a pre-assigned parameter based on the importance of different segments and the data rate available. After a segment is selected, a conventional encoding is performed over the source packets contained in that particular segment only. Thus instead of defining a UEP scheme as a set of rates (one for each segment), it is equivalently defined by a set of selection probabilities. This resembles the method of [21]. For practical reasons the number of layers in the UEP is usually constrained to two or three [92].

Note that the UEP scheme allocates redundancies to the segments based on their importance. The optimal rate allocation depends not only on the channel characteristics but also on video data since the importance and sizes of the segments vary from a GOP to another GOP. Thus, the UEP has to be dynamically changed and optimal allocation needs to be found for each GOP, which is practically feasible only for a pre-recorded video. Note that in the extreme case when the bandwidth is very scarce or packet loss rate is high, which is often the case in mobile wireless scenarios, the optimal selection probability of low-priority segments would be zero and all redundancy would be allocated to the high-priority segments to ensure their successful decoding [92].

Motivated by this and targeting wireless applications with limited bandwidth available and high loss rates, another scheme is introduced called the RASSA scheme. The RASSA scheme is a special case of UEP that exploits the flexibility of layered coding of DP and slicing. First, given an estimated packet loss rate and total rate budget, the system calculates the required overhead (and thus also the amount of source data) that will allow for error-free transfer with high probability (w.h.p.). Then, the data is filled starting from leftmost in Figure 3-1 (b), and remaining source data is discarded. This way, the scheme discards some of the lower-priority data by assigning it zero selection probability, to increase protection of the more important data [92].

Thus, this scheme is not constrained in having two or three segments/layers, and any number of DPs/slices can be selected enabling a very flexible rate control. For example, given channel statistics, enough redundancy can be provided for a segment containing DP A and B and part of DP C to be recovered at the decoder w.h.p. The unselected low-priority data (remaining DP Cs) are simply discarded. Note that the entire sent source block will be decoded, or decoding will fail, in which case the previous GOP is used for reconstruction [92].

RASSA can be seen as a UEP scheme since it protects only one part of the encoded data and discards the rest, but also as EEP since it provides equal protection of all sent source data. One immediate advantage of this scheme is reduced complexity since only one code is used, where UEP generally requires one code for each layer, and there is no need for complex rate optimization. Indeed, once the channel loss rate is estimated, the required code rate is set, and based on the available bandwidth (total budget) the decision to drop some of the NAL units that cannot fit the total budget is made [92].

UEP schemes require that the DPs of each type in LPL are aggregated together. To pass this information to the decoder, a video-table structure is proposed to be created at the encoder. The encoded video generated by the H.264/AVC encoder with DP is used to create a video-table with an entry for each NAL unit and its length. The number of NAL units per GOP is usually small (up to 64), and hence the table can conveniently be passed to the decoder within a header with negligible rate increase. The packet bearing the header will be transported with HPL. If HPL is lost then otherwise no video decoding is possible. At the receiver side, the video-table structure is used to rearrange the DPs to their original encoding order. The table is also used to discard NAL units with missing data. That is, since one DP/NAL can be sent in multiple packets, if one packet is missing the entire DP is dropped. Also, recovered DP B and DP C of a frame are dropped if DP A for that frame is not recovered properly [92].

There is negligible latency involved in bringing the DPs to their original order for decoding. The aggregation of DPs is only limited to a priority layer. For instance, if DP A and DP B both are in HPC then they will remain in their original encoding order [92].

The numbers of DPs which are dropped in RASSA scheme are only from LPC and their number is limited. Thus, it is only limited LPC data which will need to be inserted in their encoding order. This is not likely to have any latency effect. If required, the data could simply be decoded in place by the decoder.

3.2.2 Protection of Sliced AVC Video

In previous work [68], a method for segmenting sliced-AVC output into multiple segments is proposed and tested based on importance of the slices for reconstruction. For example, two priority classes can be formed where more important slices that contribute to the PSNR level above a fixed threshold are put in the high priority layer (HPL), and all other in the low priority layer (LPL). Then, the protection methods described above (EEP, UEP, and RASSA), can be applied to such prioritized data without modification [92].

A video sequence is encoded using slicing with each frame divided into a fixed number of slices. The priority of each slice is obtained by dropping it from the GOP data and measuring the resulting PSNR, as a frame-by-frame average of the entire GOP, by actual decoding. Hence this scheme is meant for pre-encoded video. This also takes into account the error propagation effect to the subsequent frames due to loss of a slice in an earlier frame. That is, the cumulative PSNR of the GOP is measured by dropping each slice in turn starting at the first P frame. After having obtained the cumulative PSNR values for each slice (as dropped), the difference from the full-decoding PSNR of the GOP is measured. The importance of the slices on total frame-averaged PSNR generally decreases going towards the end of the GOP. Thus the slices can be sorted into multiple priority layers and assigned a higher degree of protection to the important layers as compared to the layers containing less significant slices. Such layering enables a prioritized data transmission with UEP schemes [92]. Details of assigning slices to different layers can be found in [68].

3.3 Rate Allocation

In this section rate allocation optimization is discussed for the three proposed schemes. DP is assumed; however, in the same way, rate allocation can be done in case of slicing [92].

Let N be the given total rate budget expressed as the total number of packets/symbols that can be transmitted for each GOP. The video is encoded using DP H.264/AVC forming either four segments: IDR, DP-A, DP-B, and DP-C, or two classes of slices. It is assumed that each segment can be truncated arbitrarily. Let K be the total number of encoded source packets/symbols. Three schemes are considered: (i) an EEP scheme that generates N packets using all K source packets and transmit them over the network; (ii) a UEP scheme that groups source data into L importance layers starting from IDR; for example, it can have L = 4 where each of four segments forms one layer; (iii) RASSA scheme that takes first $K_{RASSA} \leq K$ source packets to generate N transmission packets [92].

Assuming that video is pre-encoded, K is fixed and is not part of the optimization. Then, EEP scheme always uses an (N,K) code and thus does not require optimization [92].

An L-layer UEP scheme can be described by L-tuples $\pi = (p_1, p_2, \dots, p_L)$ and $k = (k_1, k_2, \dots, k_L)$,where p_i and k_i represent the selection probability and the size in packets, respectively, of layer i. Then the optimal rate allocation between the L layers can be found by maximizing the expected PSNR of the reconstruction given by

$$\hat{PSNR} = \sum_{i=0}^{L} P_i \ (\pi, k) PSNR_i$$

where P_0 is probability that no layer is recovered, and P_i is probability that first i layers can be recovered but not layer $i + 1$, and P_L is the probability that all layers can be recovered successfully. The task is to find L-tuples π^* and k^* that maximize the expected PSNR, over all possible L-tuples π and k. P_i can be obtained experimentally or for some FEC codes estimated analytically for each π, k, and each channel condition and are source independent. For simplicity, it is assumed that k is set a priori by the video encoder, which is often the case. Indeed, it is natural to group all packets from one segment together. For example, for L = 3, IDR and DP A can be placed into one layer, DP B in another, and DP C in the last layer. Note that the sizes of each segment are determined by the video encoder, and are not subject of the optimization. The problem can further be simplified by maximizing the expected received rate instead of PSNR as [92]:

$$\hat{R} = \sum_{i=0}^{L} P_i(\pi) k_i$$

where k_i is the number of packets in the first i layers and $k_0 = 0$. This way, the optimization is independent of the source content and depends only on the total rate, layer sizes, and channel loss rate. There are many methods proposed to efficiently accomplish the two optimization tasks (see [61], [21] and references therein) [92].

For the RASSA scheme, recall that out of K generated source packets, only K_{RASSA} are selected that are protected by an (N, K_{RASSA}) channel code before transmission. The optimization problem is simplified to the following. Given a total number of transmission packets N and packet loss rate q, the task is to find the number of sent source packets $K_{RASSA} \leq k$, such that all K_{RASSA} source packets can be decoded w.h.p. Note that determining K_{RASSA} implies the used channel code (N, K_{RASSA}). Again, the expected PSNR or the expected number of received source packets is maximized, given by [92]:

$$\widehat{PSNR} = (1 - P)PSNR_0 + P.PSNR_1$$

and

$$\hat{R} = P.K_1$$

respectively, where P is the probability of successful decoding and $PSNR_0$ and $PSNR_1$ are reconstructed PSNR if decoding fails or is successful, respectively. K_1 denotes the number of source packets sent by the RASSA scheme. Note that P depends on q and K_1 and can be found experimentally or analytically. Indeed, for maximum distance separable codes, P is probability that the number of received packets is at least K_1, and then:

$$\hat{R} = \binom{N}{K_1} q^{N-k_1}(1-q)^{k_1} K_1$$

which can be solved numerically [92].

In the next section, results of the rate and PSNR- optimized RASSA schemes are compared to that of EEP and optimized UEP schemes.

3.4 Results and Analysis

The robustness of the EEP, optimized UEP, and RASSA schemes are tested when packet loss rates q and data rates N vary. The effectiveness of the proposed approach is shown using both DP and slicing features. Simulations have been performed using the H.264/AVC software JM 16.2 [54]. A GOP size of 16 frames is used with the IPPPP... structure [92].

3.4.1 DP AVC Transmission

It is assumed that the video has been encoded at a fixed rate using DP into fixed length segments IDR, DP A, DP B, and DP C. The data in each segment is formed into source symbols/packets of size 70 bytes for the LT coding process, which is a good compromise between performance and complexity. The IDR frame is put in the first NAL unit and it is not partitioned. CIP is used to make the decoding of DP B independent of DP C. Each non-IDR frame is partitioned into DP A, DP B, and DP C [92].

Table 3-1: Partitions for first GOP of "Paris" and "Football" sequence [92].

Partition	Paris		Football	
	Size	Cum. PSNR	Size	Cum. PSNR
IDR	22281	-	23374	-
DP A	12838	30.13	22823	-
DP B	97	30.32	2893	25.39
DP C	45732	39.16	31731	32.62
Total	80948	39.16	80821	32.62

The partitions with their relative sizes and the PSNR contribution for the first GOP of the CIF format "Paris" and "Football" video sequences are shown in Table 3-1. A two-layer UEP scheme is considered where the first, high-priority layer (HPL), contains the selected more important partitions, and the second low-priority layer (LPL) contains the remaining partitions. The UEP schemes are described by UEP ($p1, p2$), where $p1$ and $p2$ represent the selection probabilities of packets from HPL and LPL, respectively, and the optimal solution can be found as shown in Section 3.3 [92].

Table 3-2: Priority classes and LT packetization for first GOP of "Paris" sequence [92].

Class	DP	PSNR	Number of LT Packets
HPL	IDR+ DP A	30.13	502
LPL	DP B+ DP C	39.16	655
Total	80,948	39.16	1157

Table 3-3: Priority classes and LT packetization for first GOP of "Football" sequence [92].

Class	DP	PSNR	Number of LT Packets
HPL	IDR+ DP A + DP B	25.39	702
LPL	DP C	32.62	453
Total	80,821	32.62	1155

The classification of the DPs and the resulting LT packets for the "Paris" and "Football" sequences is shown in Table 3-2 and 3-3 respectively [92].

After FEC coding, one encoded symbol (together with RTP/UDP/IP headers) is placed in an IP packet and is subjected to a uniform loss pattern with average loss rates of 5, 10, 15 and 20%. Header compression is assumed, and thus a 4 byte header is considered. The base data rate is set to 1000 kbps, and the successively higher rates are obtained by adding roughly 10% additional symbols, up to a rate 1.5 times higher than the base rate [92].

The simulations are performed using one slice per frame, and a frame rate of 25 frames per second (fps). The selected schemes are simulated with 100 runs for each GOP [92].

The results for the EEP and UEP schemes are provided and compared to the results obtained with two optimized RASSA schemes: SS-PSNR and SS-Rate scheme. The results with frame-by-frame average PSNR performance of the five selected configurations at 1.1 Mbps for selected PLR are shown in Figure 3-2 and Figure 3-3, for the "Paris" and "Football" sequence, respectively. "Opt-UEP" denotes the scheme that is optimized for each packet loss rate. As can be seen from Figure 3-3, the performance of the EEP scheme is the worst. The performance of the UEP schemes gets better with an increase in the protection of HPL. UEP (60,

40) performs worse as compared to UEP (80, 20) because the protection gets divided over both segments and none is protected enough. SS-PSNR performs the best of all the schemes. For the "Football" sequence performance of the optimized UEP scheme is very close to that of the SS-PSNR [92].

The performance of all the schemes for "Football" sequence in Figure 3-3 is similar to that of "Paris" sequence in Figure 3-2. However, an overall downward trend for all schemes is seen which is due to relatively low PSNR values of "Football" sequence.

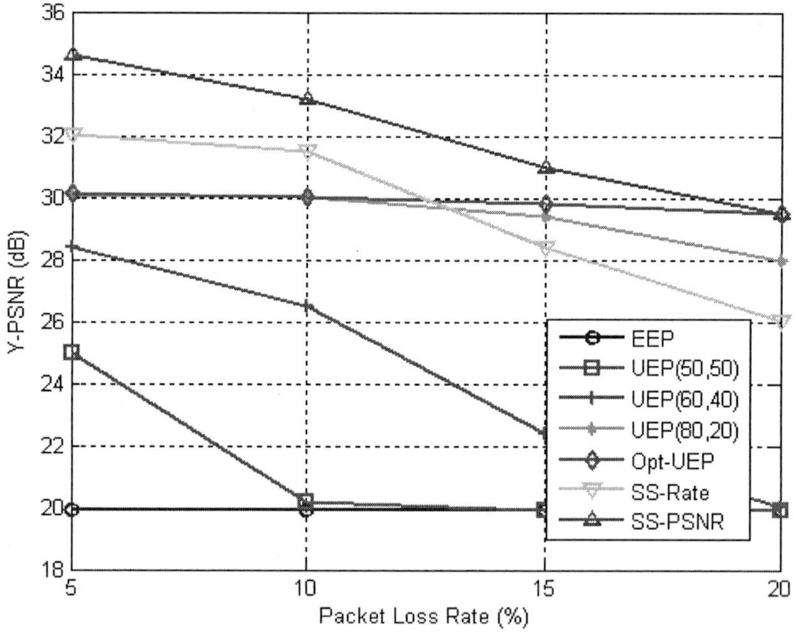

Figure 3-2: PSNR vs. PLR at overall data rate of 1.1 Mbps for Paris sequence [92].

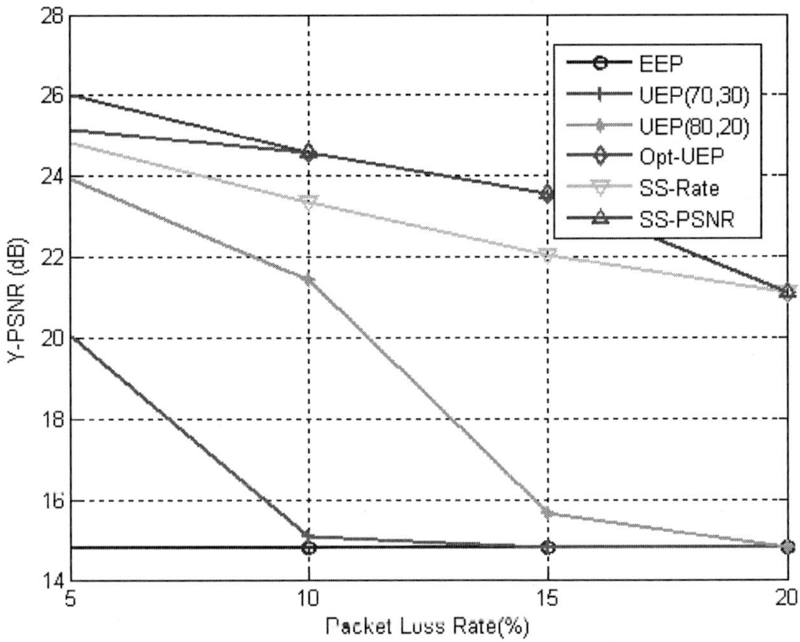

Figure 3-3: PSNR vs. PLR at overall data rate of 1.1 Mbps for Football sequence [92].

The results showing PSNR performance of the five selected configurations at 10% packet loss rates for different data rates are shown in Figs. 3-4 and 3-5, for the "Paris" and "Football" sequences, respectively. The performance of the EEP scheme gets progressively better at higher data rates. SS-PSNR and SS-Rate provide reliable and consistent performance at all the data rates. UEP (80, 20) is limited to 30 dB in Figure 3.4 even at higher rates because the DP C is not getting enough protection. Interestingly, at the highest rate the EEP scheme is better than the optimized UEP scheme, due to the absence of the performance penalty introduced by DP [92].

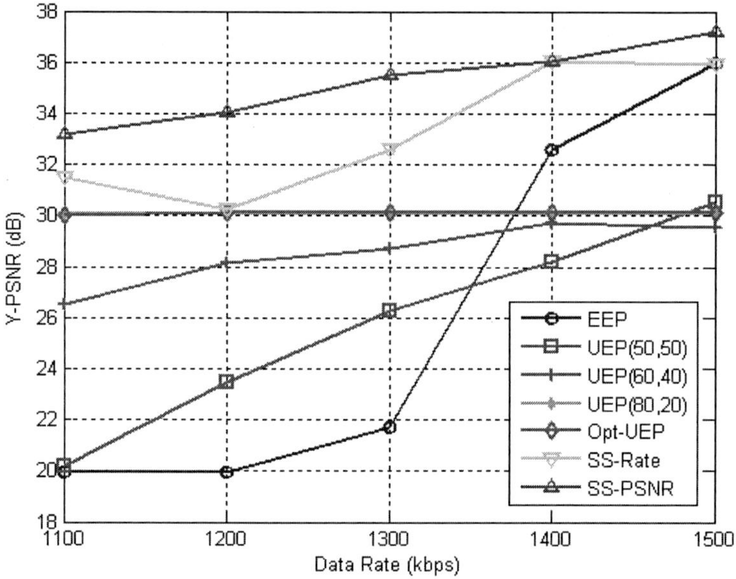

Figure 3-4: PSNR vs. data rate at packet loss rate of 10% for the "Paris" sequence [92].

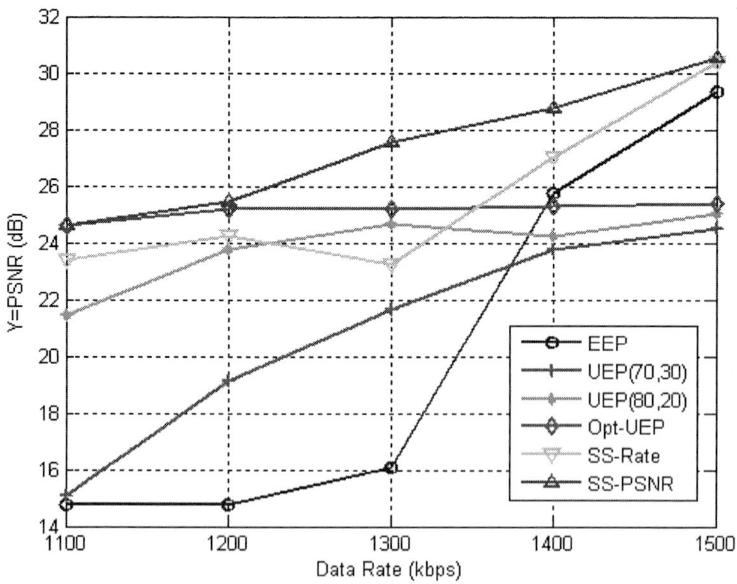

Figure 3-5: PSNR vs. data rate at packet loss rate of 10% for the "Football" sequence [92].

Frame 15 of Football sequence has been extracted for different configurations and displayed in Figure 3-6. It can be seen that the dropping of DP C as suggested in RASSA scheme does not bring much quality degradation.

(a) Original Football Sequence.
PSNR 32.62 dB.

(b) Last 4 DP C dropped.
PSNR 31.36 dB.

(c) Last 8 DP C dropped.
PSNR 29.60 dB.

(d) HPC with 15 DP C dropped.
PSNR 25.39 dB.

Figure 3-6: Football sequence with some DPs dropped.

Table 3-4: Priority classes and LT packetization for the first GOP of "Paris" sequence [92].

Class	Size(bytes)	Cum. PSNR	LT Pkts
HPL	34779	24.3	497
LPL	45536	39.08	651
Total	80315	39.08	1148

3.4.2 Sliced AVC Transmission

In this section the simulation results with the slicing feature are presented. For simplicity, the case of $L = 2$ layers is considered: HPL that contains more important slices and LPL that contains less important slices [68]. The same video parameters are used as in the previous subsection. The sizes, number of packets and resulting PSNR values for the "Paris" video sequence are shown in Table 3-4.

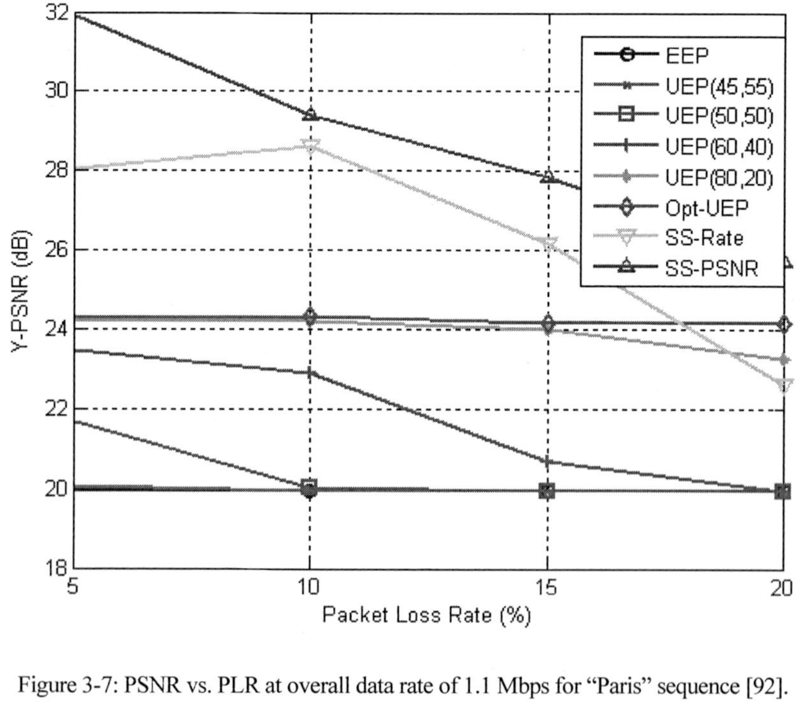

Figure 3-7: PSNR vs. PLR at overall data rate of 1.1 Mbps for "Paris" sequence [92].

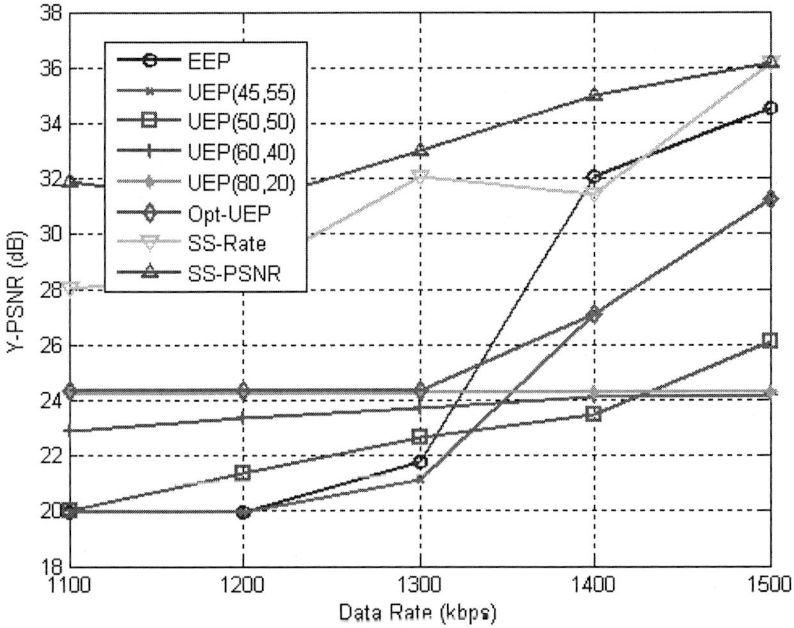

Figure 3-8: PSNR vs. data rate at packet loss rate of 10% for the "Paris" sequence [92].

The results are shown in Figures 3-7 and 3-8 and confirm the analysis carried out with the DP schemes. SS-PSNR is the best scheme overall. The UEP schemes, except UEP(45,55) in Figure 3-8 are around 24 dB as they suffer from an over protection of HPL. This is because the HPL size is only about 43% of the GOP size. This highlights the significance of considering the HPL size while designing UEP schemes. The EEP scheme becomes better than the UEP schemes at high data rates [92].

3.5 Conclusion

Although both DP and slicing have been demonstrated to enable efficient layered video data transmission, the results with DP are seen to be better. The sizes and number of DP generated are as determined by the encoder. The prioritization of data into various partitions is thus optimum and can easily be used to create different rate points. Slicing, on the other hand, is more flexible as it allows for a finer layered structure. Moreover, in contrast to DP, slicing is available in the Baseline AVC profile. However, simulation results show small advantage of the DP-based scheme compared to the slicing-based, especially at high packet loss rates [92].

The performance of different coding schemes with segmented H.264/AVC video data has been analyzed. The segmented data can be selected to suit the available data rate and channel conditions. The UEP schemes provide better performance over EEP at some rates only. The RASSA scheme can be used to match the available transmission video data to the instantaneous channel conditions. It combines the best of both the EEP and UEP schemes to provide better and reliable video quality even in the worst channel conditions. The passing of the video-table to the decoder is a low-cost solution to an "all or nothing" decoding. Note that it is assumed that the video is pre-encoded, and thus the best way to match the source rate with the channel rate is to selectively drop some of the DPs, which is done in RASSA. Indeed, the results presented here, show that the pure UEP with fixed source rate suffers huge performance loss compared to the scheme that adjusts the source rate. The main advantage of the proposed scheme is a very simple adaptation of the source rate via DP AVC coding. Note that RASSA can be combined with UEP to better match source and channel characteristics. However, that would require multiple channel codes, increased complexity, UEP optimization algorithms, and reduction of the channel code length used could worsen channel codes' correction capabilities [92].

The combined use of FEC and adaptively dropping some DP to maximize PSNR is thus shown as a practical method to ensure reliable delivery of multimedia data over wireless channels [92].

The DP and slicing features as described here are combined in Chapter 5 to split each of the slices into partitions (with DP) and hence design a novel MDC scheme.

CHAPTER FOUR

VIDEO DELIVERY OVER WIRELESS RELAY NETWORKS

4.1 Introduction

In cooperative communications, different nodes in a network collaborate with each other to transmit information from a source node to destination node over multiple parallel paths [76]. This can not only result in reduction in power consumption by the low power mobile devices, but at the same time, it also increases the reliability [93]. The cooperative communications can also increase the capacity of wireless networks. The intermediate nodes in a transmission path could act as relays of the transmitted data from the source increasing the coverage and throughput of the network. These cooperative relay networks enable a direct, single-hop and multi-hop communication to take place simultaneously from a source to the destination. RLC are ideally suited for such multiple path transmissions as the aggregated received symbols over multiple paths helps decoding at the destination [93]. In this chapter EW RLC is used for transmitting unequal error protected layered video data over a cooperative relay network. The results can be used for visual sensor and other similar networks.

Seamless and uninterrupted real-time multimedia transmission to heterogeneous mobile clients can be enabled by exploiting different diversity techniques to rectify the severe noise and fading effects. Multipath diversity together with MDC [69], [70] efficiently combats noise, fading and shadowing in mobile channels [94] © 2013 IEEE.

Emerging wireless communications systems, such as LTE-A Release 10 [71], [72] and mobile WiMax, offer the possibility of increasing system capacity and extending coverage via relaying and cooperation [26]. In LTE-A, for example, the relays receive and retransmit the signals between the base station and mobiles, resulting in effective increase in the throughput and extending the coverage of cellular networks [73]. In order to autonomously and dynamically adapt to network bandwidth, delay, and channel noise fluctuations, a promising solution is to combine multipath relaying with MDC [69], [74]. MDC generates multiple descriptions of the same source data that can be independently decoded and is very well suited to emerging relay-assisted networks, where each description can be routed via a different relay exploiting network diversity. However, since mobile channels are very prone to packet losses, FEC is still necessary to ensure that enough descriptions are correctly received [94] © 2013 IEEE.

For relay based configurations the paths have been assumed to be symmetric with reference to the delay encountered by packets traversing each path. Thus, it is assumed that packets transported over multiple paths which converge at a destination, each of them arrive so as to meet the delay deadline, unless lost based on the channel model used.

AL-FEC provides adaptive resilience to packet losses, which are frequent in mobile channels. Recently, AL-FEC via Raptor codes [9] has been standardized for DVB-H and MBMS services [35]. RLC [13], [36], [20], like Raptor codes, are rateless, capacity approaching, and low encoding complexity erasure codes. However, with network coding principles at its roots (see [13] and references therein), RLC enable simple and efficient relaying and network cooperation [94] © 2013 IEEE.

A novel and simple MDC scheme has been designed. In the proposed scheme independent descriptions are created with least possible duplication using slicing and DP [2]. Although slicing and data partitioned video data are used to apply EW-RLC for error protection, the proposed scheme can be considered for other MDC schemes e.g., as in [77], redundant slices are used for creating multiple descriptions.

A system is proposed for real-time video streaming to mobile users, via LTE-A and similar relaying wireless systems, using H.264 AVC [2] based layered MDC together with RLC. The source data is encoded in multiple descriptions, which are sent to a receiver over non-cooperative relays. The receiver can reconstruct the encoded data from any subset of the descriptions received. In order to maximize benefits of relaying and MDC, two algorithms are proposed for relay selection and optimal resource allocation that minimize reconstructed source distortion [94] © 2013 IEEE.

In [82], an amplify-and-forward wireless relay system is considered, and novel power allocation strategies are proposed, based on geometric programming, that optimize the maximum transmit power and the network throughput. In [83] a dynamic resource allocation of transmission powers and sub-channels under power constraints based on instantaneous channel states that maximize the instantaneous total transmission rate is investigated and a greedy approach proposed. In [84], a globally optimal Pareto-based solution and sequential optimization algorithm using channel state information are proposed for the OFDMA-based half-duplex single-relay channel. In contrast to the above papers which provide "lower layers" (physical, link, and/or network layer) parameters selection, this work focuses on the application-aware resource allocation, where upper-layer parameters (such as AL FEC) are considered as well, and the reconstructed source fidelity is used as performance measure [94] © 2013 IEEE.

The second category of related work covers path selection and resource allocation for multipath video. In [85], optimal and polynomial-time suboptimal algorithms for selecting the network path and scheduling packets of encoded scalable video are proposed that minimize video distortion and time delay at the receiver. In [86], MDC relaying is considered, and an adaptive compression scheme is proposed at the source and relay based on the received feedback. In [78] a joint source-channel scheme is proposed that exploits information about packet losses to adaptively select reference frames in MDC. In [79] FEC and routing are jointly optimized to maximize reconstructed quality. The main difference between this chapter and the above contributions is that the proposed resource allocation strategy considers both lower and upper layers parameters and proposes a dynamic programming and a suboptimal greedy solution. In contrast to [78], [79], video encoding is not altered, making it compatible with the standard, but only the coding rate is adapted, and relay selection is done based on the channel characteristics [94] © 2013 IEEE.

Targeting applications such as [80], LTE-A network is modelled using a Finite-state Markov chain (FSMC) [81] path loss channel model using LTE-A Release 10 parameters (suggested in [75]) [94] © 2013 IEEE.

The rest of the chapter is structured as follows. Multi hop cooperative relay networks comparing results for AF and DF schemes are described for a generic system in Section 4.2. The design of an adaptive layered MDC coding scheme for wireless video is described and the scheme is simulated for a technology independent multi-path model in Section 4.3. The next three sections specifically address relay based communication as envisaged in LTE-A. The proposed system for relay-assisted rateless layered MDC is described in Section 4.4. The resource allocation optimization is presented in Section 4.5. The simulation setup and results are provided in Section 4.6. Finally, Section 4.7 concludes the chapter.

4.2 Multi hop Cooperative Relay Networks

The focus of this section is to compare the performance of layered video data over multi-hop relay networks using the EW approach [21] with RLC for video streaming. The layered video data is obtained with DP feature of H.264/AVC and quality scalability with H.264/SVC.

The simulation results show that EW RLC can be used effectively for reliable video transmission over the emerging relay networks. The results can easily be extended to more than one relay node. Although schemes exist which allow for passing back of an acknowledgement [87] of video decoding at the relay or the destination, this analysis is restricted to the case without provision of a feedback.

In this section, Amplify-and-forward (AF) [88] and Decode-and-forward (DF) schemes are analyzed for streaming of layered video with EW-RLC over relay networks. An adaptive scheme is proposed to vary the degree of protection to the most important layer to improve the overall quality of received video.

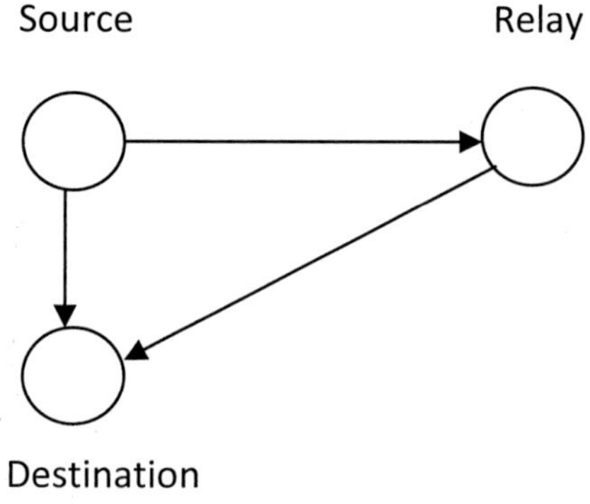

Figure 4-1: 3-relay node model [93].

4.2.1 System Description

The nodes in multi-hop relay network typically communicate over unreliable wireless channels and have limited transmission power. The analysis uses a simple 3-node relay model which comprises of a source (S), relay (R), and a destination (D) node as shown in Figure 4-1. It is assumed that the destination is able to simultaneously receive transmissions from S and R and that the relay is half-duplex that is it cannot transmit and receive at the same time. The channel from source-to-relay (SR) and relay-to-destination (RD) is assumed to be better as compared to the direct channel between source and destination (SD). If the SD channel has low loss, then there may be no requirement for relay collaboration. However, if the direct channel SD experiences loss then the relay may enter the collaboration state, where it tries to improve the received video quality by additional decoding of data received via the relay at the destination node [93].

The relay collaboration could use an AF scheme [88] where the relay acts as a repeater, simply re-transmitting each received packet towards destination node. Note that the same code is being transmitted on the channel SD; hence if both packets are received by the destination node, it does not bring any decoding advantage. At the destination the decoding is attempted at the end of transmission [93].

For the other relay collaboration scheme using DF, the relay continues to accumulate the packets until such time that the received data is decodable. As soon as the relay completes decoding, it re-encodes the video data and starts transmitting. Early decoding at the relay is advantageous since it reduces end-to-end delivery when poor channel conditions result in loss on the direct link, SD, from the source to destination [93].

4.2.2 Measurement Setup

The well-known video sequences Paris and Football were used for the simulations. This choice was made because the Paris sequence has a high spatial complexity and the Football sequence has a high temporal complexity [93].

Both sequences were in CIF format with 25 frames per second (fps), and the CIP flag set. The GOP size was 16 frames, with an IPPP... structure. The total number of video packets for each video configuration was about the same to enable a comparison of the schemes [93].

The video was packetized into packets of 1024 bytes. A packet header of 6 bytes was assumed, anticipating the use of packet header compression, common on wireless networks. Each simulation was repeated 1000 times and the results averaged [93].

A set of simulations was performed using EW-RLC to compare the performance of AF and DF for transmission over the relay model. A Finite-state Markov chain (FSMC) [81] channel model was used to model the network links [93].

The equal loss protection (ELP) scheme, with no priority for HPL is compared to the unequal loss protection (ULP) schemes. For ULP schemes, the probability of selection (PS) for HPL was varied among the values of 0.50, 0.60, 0.80 and 1.0. This created four ULP schemes. With a PS of 0.60, the symbols were generated from the HPL with a 60% probability, hence, affording better protection to HPL. The size of the HPL also governs selection of an appropriate PS. The PS should be higher for ULP compared to the size (percentage) of HPL in comparison with the LPL.

The simulations considered the case where there was a better channel condition for the relay than using the direct path. The packet loss rate (PLR) for the RD channel was constant at 0.025. The PLR for the direct SD channel was varied within the set: 0.05, 0.10, 0.20, 0.30, 0.40, and 0.50 [93].

Loss concealment is common for packet video receivers. We considered a simple method that replaces a missing frame (or an entirely missing GOP) with the last frame previously decoded [93].

Table 4-1: Relative Partition Sizes – Paris Sequence [93]

Partition	Size (bytes)	Size (%)	Cumulative PSNR
IDR	22,281	27.52	-
DP-A	12,838	15.86	-
DP-B	97	0.12	30.32
DP-C	45,732	56.50	39.16
Total	80,948	100	39.16

Table 4-2: Relative Partition Sizes – Football Sequence [93]

Partition	Size (bytes)	Size (%)	Cumulative PSNR
IDR	23,374	28.92	-
DP-A	22,823	28.24	-
DP-B	2893	3.58	25.39
DP-C	31,731	39.26	32.62
Total	80,821	100	32.62

Table 4-3: Packetization of Layers [93].

Sequence	Number of Packets		
	HPL	LPL	Total
Paris	35	45	80
Football	49	31	80

4.2.3 Video Transmission with H.264/AVC

4.2.3.1 System Configuration

The DP feature was used to create encoded data with partitions. These partitions were used to form two layers of packet video that were protected with ULP using EW-RLC [93].

The breakdown of the video into constituent partitions for the Paris sequence and Football sequence are respectively shown in Table 4-1 and 4-2. After encoding, the IDR frame together with DP A and B were placed in High priority layer (HPL), and the remaining portion, that is, DP C was placed in Low priority layer (LPL) [93].

The packetization details for each sequence are given in Table 4-3. This table shows that the number of encoded packets for the HPL using the Football sequence is substantially higher compared to the Paris sequence. This implies that the same ULP strategy does not perform equally when encoding both sequences, demonstrating that performance is linked to video content structure [93].

4.2.3.2 Simulation Results

The results for the Paris sequence are shown in Figure 4-2. The scheme AF50 denotes an AF scheme with a PS of 0.5. The notation AF-ELP denotes protection over the entire video data, treating all data with the same priority. The results show that the performance of ELP scheme is much lower than when using ULP. DF-ELP has the poorest performance, because it is unable to decode using the relay, and hence preventing relay collaboration [93].

Two notable cases are AF100 and DF100. These exactly overlap, but provide a consistent PSNR contribution of 30.32 dB irrespective of the PLR. With an over-protection for HPL the performance is limited to PSNR achieved by only receiving with the HPL (see Table 4-1) [93].

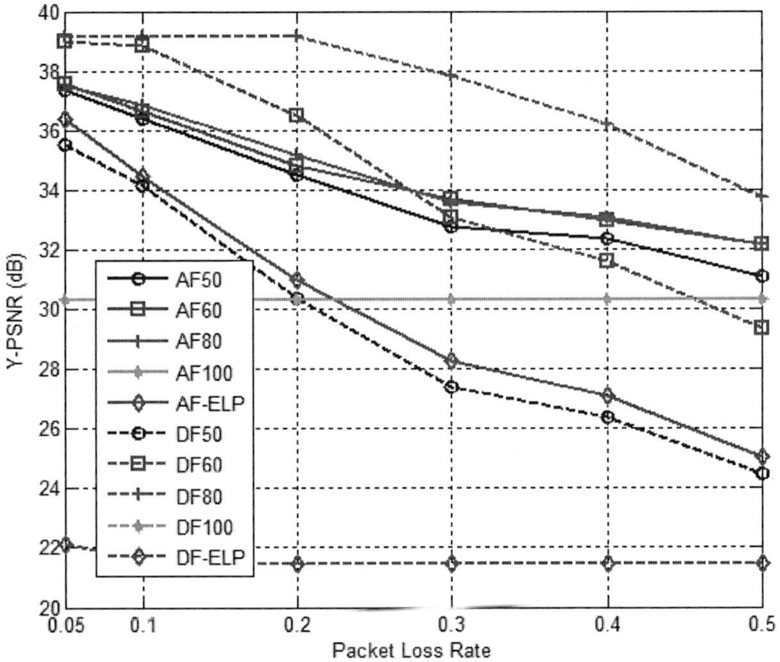

Figure 4-2: PSNR vs. PLR for Paris sequence (AVC) [93].

The performance of DF improves with an increase in PS of the HPL, such that DF80 provides significantly better performance. This is attributed to the higher protection offered by the HPL, the probability of an early decoding of HPL at the relay node increases, resulting in an early start of the relay collaboration, reducing time for decoding at the destination [93].

The results for the Football sequence are shown in Figure 4-3. The performance of DF schemes was generally reduced at a low PS, where the HPL was much smaller than the LPL in the Football sequence; hence successful decoding of HPL was delayed using the DF scheme. This resulted in a sharp decrease in the performance at a high PLR, with no relay participation, especially when using a DF scheme with a low PS. The AF100 and DF100 were able to maintain performance across the range of PLR [93].

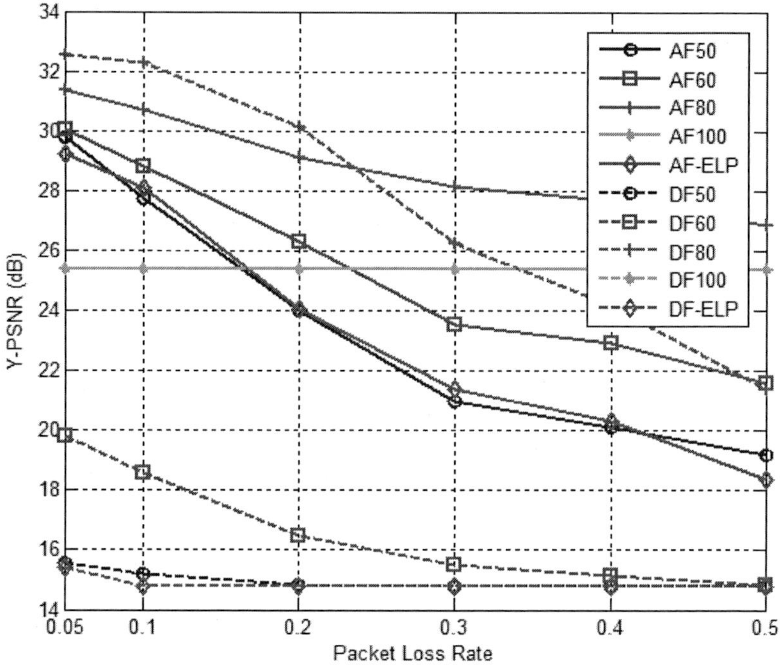

Figure 4-3: PSNR vs. PLR for Football sequence (AVC) [93].

4.2.4 Video Transmission with H.264/SVC

4.2.4.1 System Configuration

Coarse Grain Scalability (CGS) was used to create scalable video. Two layers of video data are created, which could then be protected by providing unequal protection by using EW-RLC. A Quantization parameter (QP) was used to tailor the size of the base and enhancement layers. This also affects the resulting achieved PSNR for each layer [93].

The breakdown of video data into the two layers for the Paris and Football sequence are respectively shown in Table 4-4 and 4-5. The packetization details are shown in Table 4-6 [93].

Table 4-4: Relative Layer Sizes - Paris sequence [93].

Layer	Size (bytes)	Size (%)	Cumulative PSNR
HPL	29,738	36.64	32.34
LPL	51,441	63.36	32.34
Total	81,179	100	37.14

Table 4-5: Relative Partition Sizes - Football sequence [93].

Layer	Size (bytes)	Size (%)	Cumulative PSNR
HPL	34,814	43.44	27.91
LPL	45,337	56.56	30.32
Total	80,151	100	30.32

Table 4-6: Packetization of Layers [93].

Sequence	Number of Packets		
	HPL	LPL	Total
Paris	30	51	81
Football	34	45	79

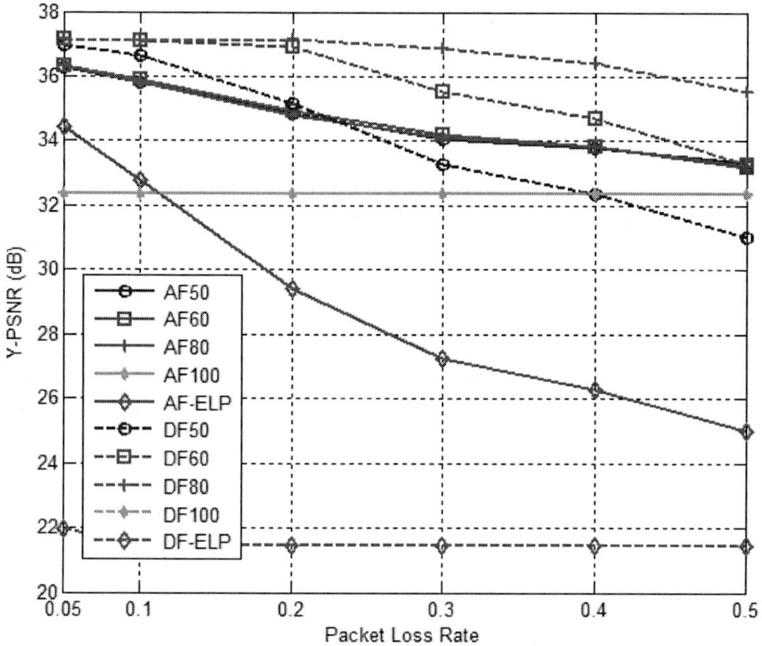

Figure 4-4: PSNR vs. PLR for Paris sequence (SVC) [93].

4.2.4.2 Simulation Results

The results for the Paris sequence are shown in Figure 4-4, showing that the performance of ELP scheme is poorer than the ULP schemes. The schemes AF100 and DF100 maintain a consistent PSNR contribution of 32.34 dB irrespective of the PLR, which is the same as the PSNR achievable using only the HPL. The performance of the DF schemes is was generally better than using AF. However, as previously the performance of DF schemes improved at a higher PS for the HPL. Again DF80 provides the best overall performance [93].

The results for the Football sequence are shown in Figure 4-5. These results confirm the results obtained with Paris sequence. However, the performance of DF schemes has slightly poorer performance than for the Paris sequence. This degradation is attributed to the larger HPL [93].

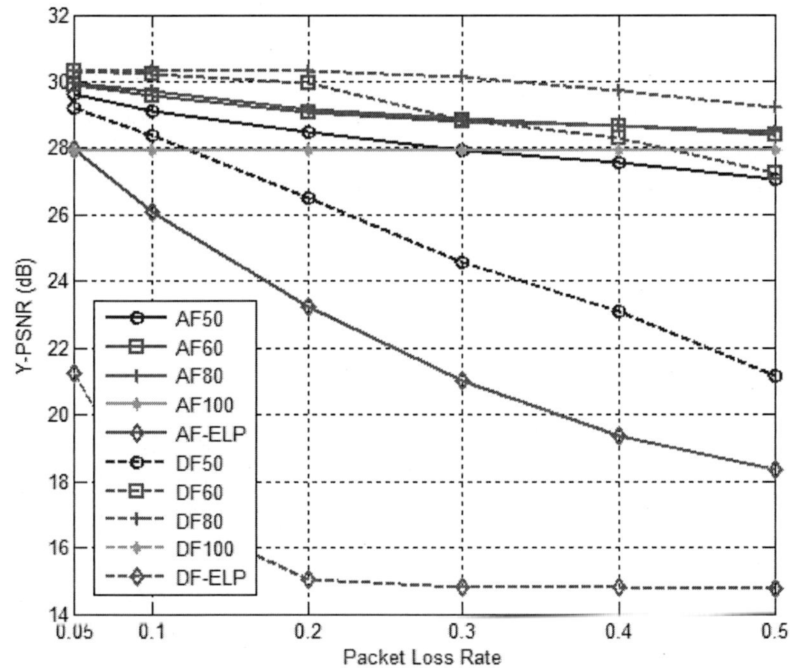

Figure 4-5: PSNR vs. PLR for Football sequence (SVC) [93].

4.2.5 Comparison and Analysis

The performance of H264/AVC and H.264/SVC can be compared using Figure 4-2 and 4-4. However, the difference in HPL sizes must also be considered. Table 4-1 and 4-4 show that the size of HPL for H.264/AVC is larger with a lower PSNR compared to H.264/SVC. H.264/SVC also imposes a loss in coding efficiency, so the total PSNR for H.264/AVC is slightly better than when using H.264/SVC [93].

An interesting effect arises when comparing AF and DF schemes with increasing PLR. This results from the contribution of the HPL compared to combined decoding of both HPL and LPL. This effect is observed for AF60 and DF60 in Figure 4-1, where the value of DF60 drops below that for AF60 at higher PLR. The effect is also seen in Figure 4-2 and 4-4, i.e., the results for DF80 for H.264/AVC are better at low PLR, but the performance of H.264/SVC is better for a higher PLR. Similar analysis was observed using AF. The conclusions are also substantiated with the results for the Football sequence [93].

The effect of HPL size on the performance of the AF and DF schemes is discussed next. The HPL size is chosen to be 30 and 70 (out of a total of 100) video packets for each of the two considered configurations. To quantify the effect, Probability of Error (PE) was kept as 0.3; while the PS was increased in steps and the resulting performance of AF and DF schemes was measured as summarised in Table 4-7. The decoding (%) implies a count of the cases when for instance only the HPL is decoded, or the GE fails to recover any useful FEC packets (the column titled failure). Figure 4-5 shows that a larger HPL size adversely impacted the performance of the DF scheme [93].

The performance of DF was better as the PS increased (Table 4-7). Hence, the PS used for the HPL needs to be selected based on the size of HPL in the video GOP data. This effect can similarly be seen when using the AF scheme. It is possible that future systems can select the PS using an adaptive scheme that assigns the PS depending upon the size of the HPL (for each GOP) increasing the video quality throughout. DF80 offered the best choice over the entire range of PSNR in a scenario where no feedback mechanism was available [93].

Table 4-7: Effect of PS on Layer Decoding [93].

HPL Size (%)	**PS (%)**	**Decoding (%), (PE = 0.3)**					
		HPL		**HPL+LPL**		**Failure**	
		AF	**DF**	**AF**	**DF**	**AF**	**DF**
30	30	43	38	21	4	36	58
	40	78	60	21	37	1	3
	50	81	17	19	83	0	0
70	70	37	2	22	0	41	98
	80	76	47	23	1	1	52
	90	77	91	23	7	0	2

The advantage of H.264/AVC is that once the DP have been created, it is possible to selectively drop selected partitions to yield a base layer of required rate for transmission [93].

For further study of relay based communication H.264/AVC standard is used. The DP feature as used in this section is used in combination with slicing to yield multiple descriptions. The design of the layered MDC scheme and simulations over a generic two-path model are provided in the next section.

4.3 Adaptive Layered MDC Coding for Wireless Video

In this section an MDC scheme is proposed based on DP and slicing feature of H.264/AVC. In the proposed scheme independent descriptions are created with least duplication and additionally EW-RLC [52] is proposed for FEC. The advantage being that the degree of protection can be adapted to suit a particular wireless channel [89] © 2012 IEEE.

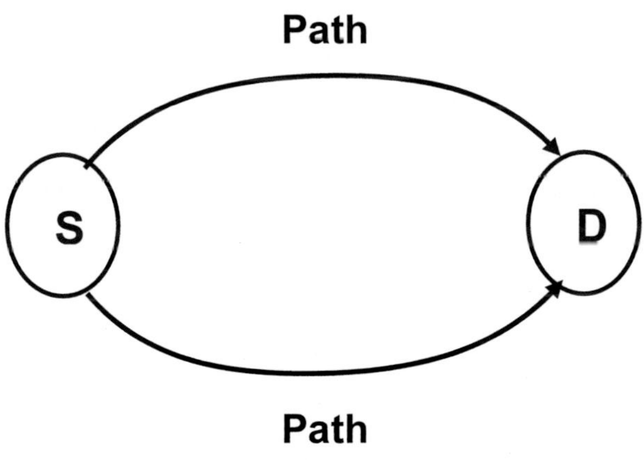

Figure 4-6: The system model.

The focus of this section is to simulate the transmission of layered descriptions over a two-path wireless network as shown in Figure 4-6 and find the best rate distribution for the descriptions that maximizes PSNR. In this section it is assumed that two independent wireless channels connect a source S to the destination D, which could exist in a multi-interface receiver in a multi-hop wireless network. The system parameters relating to the channels are the data rate and packet erasure rates [89] © 2012 IEEE. The possible applications are in multi-interface networks and VoD applications.

4.3.1 System Description

Two independent descriptions with equal size and importance can be created by dividing the slices of a frame into two sets. As shown in Figure 4-7, each frame has been divided into six slices, and each description comprises of three slices from each frame. The decoder is thus able to create the missing information in case of losing any description [89] © 2012 IEEE.

In order to make use of EW-RLC codes over each description, the DP within each slice can be assigned to different layers for unequal protection [89] © 2012 IEEE.

The simulations have been performed using Foreman sequence in CIF format. The CIP flag has been set and MBIU are used to contain the error propagation. The GOP size is 8, and 6 slices per frame are used. The video data rate is 1000 kbps for 25 frames per sec. Each simulation is run 1000 times for the results to converge. Two descriptions using slice distribution are created for use in all the simulations. Each of the description contains 3 slices from each frame as shown in Figure 4-7. Note that each of the descriptions has the IDR frame to make possible the decoding of each description [89] © 2012 IEEE.

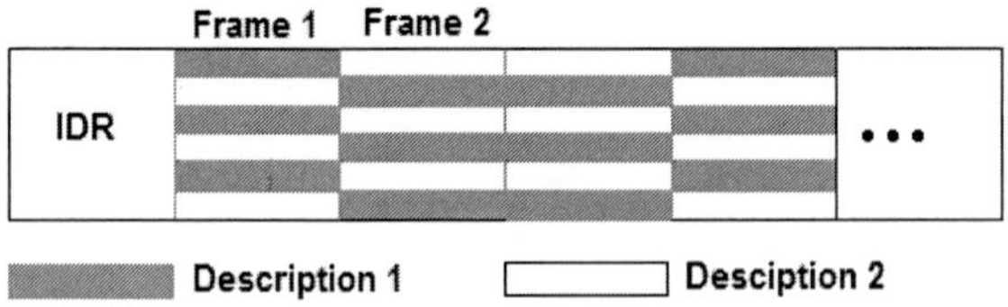

Figure 4-7: Slice layout for creating descriptions. Each rectangle represents a slice [89] © 2012 IEEE, [94] © 2013 IEEE.

In order to enable unequal protection to each of the description's data, two importance layers are identified. The data within each of the slice is already partitioned into DP A, B and C. The most important data for video re-construction, consisting of IDR, DP A and DP B is termed as high-priority layer (HPL). The remaining data i.e. DP C constitutes low-priority layer (LPL) [89] © 2012 IEEE.

Table 4-8: System configuration [89] © 2012 IEEE.

Cat.	GOP		MDC1		MDC2	
	Size (bytes)	Cum. PSNR	Size (bytes)	Cum. PSNR	Size (bytes)	Cum. PSNR
IDR	11515	-	11515	-	11515	-
DP A	9922	32.45	4892	28.37	5030	27.92
DP B	4352	33.91	1744	28.76	2608	28.49
DP C	14828	40.1	7538	29.55	7290	28.92
Total	40617	40.1	25689	29.55	26443	28.92

Table 4-9: PSNR Contribution [89] © 2012 IEEE.

Categories Received		PSNR (dB)
MDC1	MDC2	40.1
MDC1	HPL2	36.27
MDC1	Fail2	29.55
HPL1	Fail2	28.76
HPL1	MDC2	36.6
Fail1	MDC2	28.92
Fail1	HPL2	28.49
HPL1	HPL2	33.91
Fail1	Fail2	20.71

The sizes of the different partitions for the first GOP of the Foreman sequence together with the assignment of the partitions to MDC1 and MDC2 are shown in Table 4-8 [89] © 2012 IEEE.

Table 4-10: Packet allocation to different layers [89] © 2012 IEEE.

Layer	MDC1			MDC2		
	Size (bytes)	PSNR	Pkts	Size (bytes)	PSNR	Pkts
HPL	18151	28.76	36	19153	28.49	38
LPL	7538	29.55	15	7290	28.92	15
Total	25689	29.55	51	26443	28.92	53

The PSNR contribution for each of the possible combinations has been calculated and is listed in Table 4-9. As can be seen from the table, the best video quality is achievable when both the descriptions are received correctly. However, the novel way that the descriptions have been designed, the video can be decoded with just the HPL of any one description only. This layering of video data within a description makes this scheme adaptive and practical for varying channel conditions. In case where both descriptions are lost entirely and nothing is decidable for the GOP then the decoder applies simple error concealment technique by replacing the lost GOP by the last frame of the previously successfully decoded GOP, which comes to 20.71 dB [89] © 2012 IEEE.

The video data belonging to each description is divided into packets of 512 bytes. The details of the packetization are as shown in Table 4-10. The simulations are carried out with both uniform loss and fading channel model to represent wireless channels [89] © 2012 IEEE.

For the video transmission, depending on the bandwidth and packet erasure rate, probability of selection, PS1 could be varied to maximize the PSNR. An interesting scenario is with PS1 as 100%, thus effectively sending HPL only [89] © 2012 IEEE.

In the absence of any errors the PSNR achieved with both the descriptions is the same as that of standard coding, considering slicing and data partitioning [89] © 2012 IEEE.

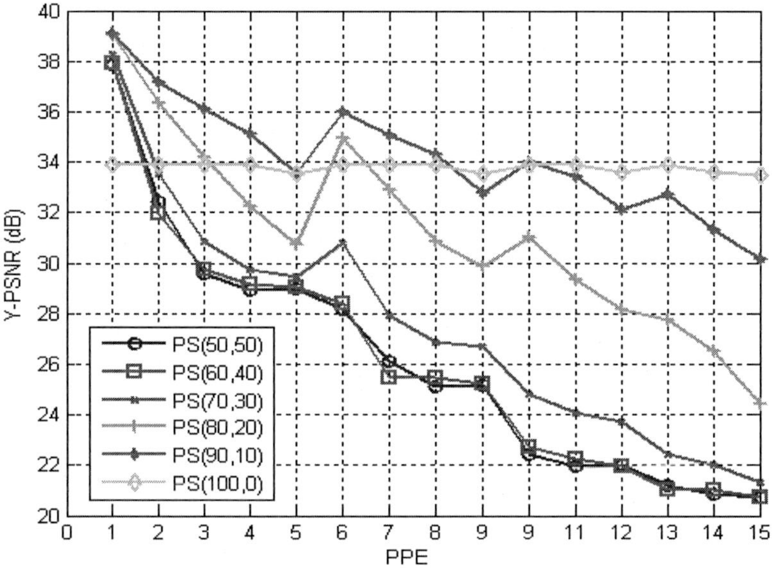

Figure 4-8: PSNR at different PPE values of Section 4.3.2 [89] © 2012 IEEE.

4.3.2 Results for Uniform Loss Model

In this section the analysis is performed with uniform loss model. The probabilities PS1 and PS2 are kept as same for both paths. The data rate for each of the paths is kept as 10% over the source rate. The probability of error (PE) on paths are varied in increments of 0.05 from 0.05 to 0.25, to yield 15 combinations (0.05, 0.05), (0.05, 0.10), (0.05, 0.15), (0.05, 0.20), (0.05, 0.25), (0.10, 0.10), (0.10, 0.15), (0.10, 0.20), (0.10, 0.25), (0.15, 0.15), (0.15, 0.20), (0.15, 0.25), (0.20, 0.20), (0.20, 0.25), and (0.25, 0.25). This pair of PE is termed as PPE1 through PPE15 [89] © 2012 IEEE.

The results for each set of PE with different PS1 are shown in Figure 4-8. Each curve (from left to right) depicts a PPE from (0.05, 0.05) to (0.25, 0.25) and shows a general downward trend subject to cumulative PE on both channels. The scheme with PS (100, 0) is seen to be virtually independent of the increasing packet erasures. Thus, by increasing the protection of HPL for a description it is possible to offset the effect of higher packet loss rates, and hence, a consistent reconstruction quality can be obtained [89] © 2012 IEEE.

The results for some PE are shown in Figure 4-9. It can be seen that as the PS is increased, all the schemes converge at around 33.91 dB which is the PSNR achieved with receiving HPL only for both MDC1 and MDC2 [89] © 2012 IEEE.

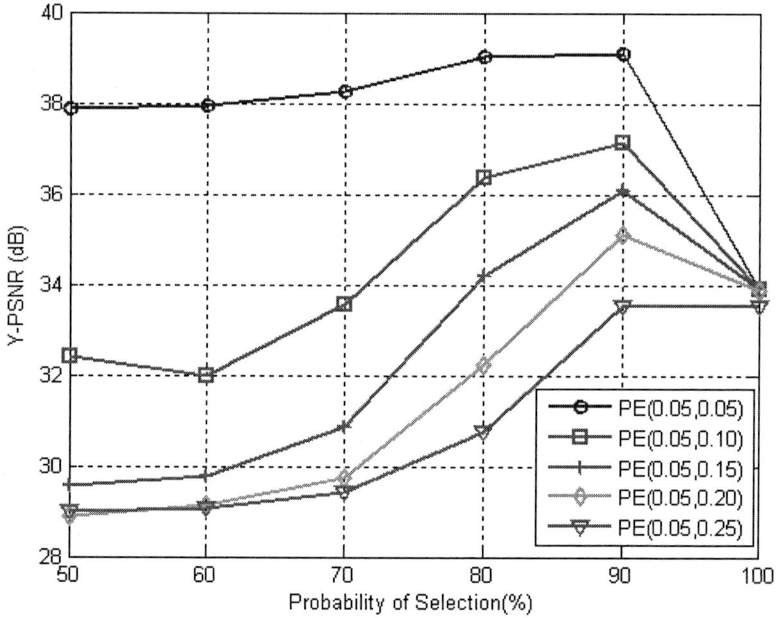

Figure 4-9: PSNR at different PE [89] © 2012 IEEE.

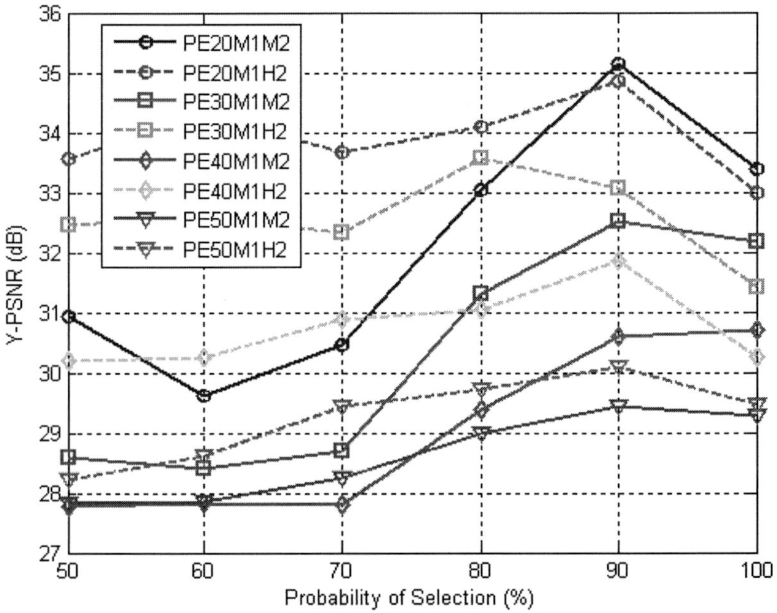

Figure 4-10: Comparison of sending (MDC1, MDC2) against (MDC1, HPL2) [89] © 2012 IEEE.

4.3.3 Results for Fading Error Model

In this section, one of the channels C1 is assumed to be better than the other C2. The PE1 is kept as 0.05, whereas PE2 is varied as 0.2, 0.3, 0.4 and 0.5. Gilbert model is used to simulate the burst losses with an average burst length of 5 packets for PE1 and PE2. The PS is varied from 0.5 to 1.0 as in the case of uniform loss. The data rate for each of the paths is kept as 10% over the source rate [89] © 2012 IEEE.

Also, to establish the importance of adapting the description to the channel bandwidth and error characteristics, we assume for C1 same configuration as described above in this section. However, HPL only is transmitted over C2. The reason for this is that for a low bandwidth channel like C2 here, the protection of whole of the description MDC2 is not possible. Results for the adaptive scheme are presented by varying the coding parameters, i.e., PS, and the information (complete description or HPL) [89] © 2012 IEEE.

The results for transmission comparing the PSNR with burst loss, for transmitting each of the descriptions, MDC1 and MDC2 in full, with transmitting MDC1 with only HPL2 at different PS are shown in Figure 4-10. The configuration PE20M1H2 is a scheme with 20% burst errors wherein MDC1 and HPL2 are being transmitted. It can be seen that the schemes with only HPL2 being transmitted have better results for the entire range of PS. Thus, in any such scenario, where the channel is incapable of successfully communicating the whole of source data, it is advantageous to reduce the source data itself. This reduction brings a drop in PSNR which is offset by gain in successful reception (from reduced information), hence providing an overall improved result [89] © 2012 IEEE.

The design of the two-path MDC scheme has been evaluated under various channel conditions. The scheme is fully compatible with the standard H.264/AVC encoder. The proposed scheme of UEP with slicing and DP of H.264/AVC constructs layered video data amenable to FEC with EW-RLC. The transmission of HPL layer only has been shown to offset the worst channel conditions. This makes such adaptation both beneficial and necessary for low bandwidth channels. This optimal scheme for rate adaptation over multiple paths for wireless video transmission can adaptively operate at GOP level [89] © 2012 IEEE.

The layered MDC scheme designed in this section with H.264/AVC is used in the next section to create layered MDC video for transmission over relay based channels simulating the emerging LTE-A standard.

4.4 System Description for Relay-assisted Rateless Layered MDC

A source node generates source blocks that need to be streamed in real time to a mobile destination. The source node independently compresses each data block of length B symbols (e.g., bytes) into D descriptions, D << B, using an MDC scheme, and each description consists of l layers (e.g., quality, temporal, etc). After source coding the descriptions are fed into a rateless AL-FEC encoder, which for each description, generates potentially infinite stream of coded packets of L symbols each. Each encoded packet is a random linear combination of the source packets over sufficiently large field. The source can send the encoded packets to the destination via a direct link, or via $N-1$ available relays. The jth channel ($j = 1,...N$) is characterized by the total number of packets that can be transmitted until the deadline, which is denoted by R_j and depends on the channel bandwidth, and the packet loss probability, p_j. Note that R_j and p_j can in general be functions of time, where p_j also varies with the relay/source transmit power [94] © 2013 IEEE.

Note that, though the source exploits rateless AL-FEC coding, due to real-time playout constraints, only a finite number of encoded packets R_j can be sent through channel j for each encoded source block. The destination collects rateless packets from all links, until it either receives enough to decode all descriptions or until playout deadline is reached. Each description can be decoded independently, the descriptions are of unequal importance to the final recovery, and each encoded description can only improve the reconstruction [94] © 2013 IEEE.

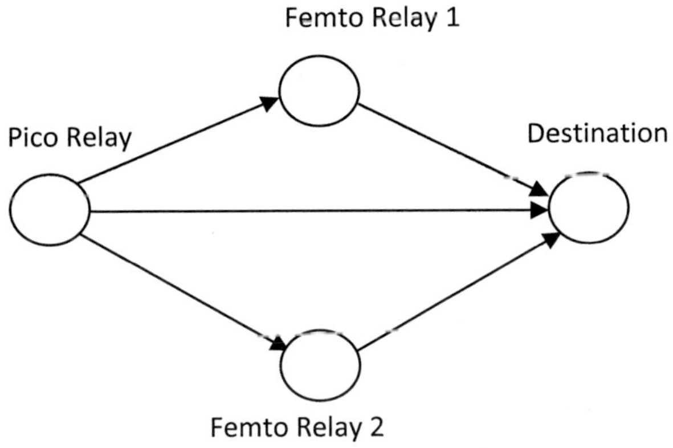

Figure 4-11: The block diagram of the system with two relays.

In particular, the above system can be used to model video streaming over LTE-A networks, where a base-station (source) in the pico-cell streams video to a mobile client via direct link and also via femto-cell relays. [94] © 2013 IEEE. A system with two femto-cell relays is shown in Figure 4-11. LTE-A transmission enables a more efficient operation of the Multimedia Broadcast and Multicast Service (MBMS) [12], allowing over-the-air combining of multi-cell transmissions towards the User Equipment (UE). Therefore a handheld mobile device can be served by multiple relays.

Due to the strict time constraints of real-time streaming, it is assumed that the employed relays do not decode the received packets but operate in the amplify-and-forward mode. As in LTE, the pico-relay or the base station (BS) uses separate orthogonal links to send packets to the relays and destination. The relays also forward the packets to the client using orthogonal channels. Since the BS and the relays are static (and these connections high-bandwidth fibre optic channels), in our analysis, it is assumed that the source-relay channels are always error-free [94] © 2013 IEEE.

4.5 Resource Allocation

Given statistical knowledge of the source content and channel characteristics, the task is to find source-channel-relay resource allocation that maximizes the reconstructed video quality [94] © 2013 IEEE.

Let $\delta \subseteq \{1,\ldots,D\}$be a subset of descriptions. Let $\pi = \{r_{ij}\}$ be a *D X N* matrix of the rate allocation strategy over N channels, where r_{ij} , $i = 1,\ldots,D$ and $j = 1,\ldots,N$, is the rate (in terms of the number of encoded packets) of description i that will be sent over channel j. Note that the total number of encoded packets of description i is $\sum_j r_{ij} = \left\lceil \frac{q_{s_i} q_{c_i} B}{L} \right\rceil$, where q_{s_i} and q_{c_i} are source and channel coding rate for description *i*, respectively [94] © 2013 IEEE.

Matrix π is thus used at the source to select how many encoded packets from each description should be routed over each relay. However, since each description consists of *l* quality layers, encoded packets are not of equal importance. Thus, furthermore, for each description, the source probabilistically schedules encoded packets for transmission using channel coding probability matrix Q_{EW} (which will be explained later, see also [22]). Thus, Q_{EW} is used by the source to decide which particular packets for each description to send. This decision is based on the importance of each packet to recovery of the description or on the time playout information and channel characteristics [94] © 2013 IEEE.

Let $\rho = [\rho_{1,\ldots}\rho_N]$ be the set of transmitter powers, where ρ_1 is the transmitter power at the source, and ρ_i , $i = 2,\ldots N$ is the transmitter power at the *i*-th relay. Then, the resource allocation problem that maximizes the total reward w is:

$$\max_{\pi,\rho,q_{s_i},Q_{EW}} w = \max_{\pi,\rho,q_{s_i},Q_{EW}} \sum_{\delta} P(\delta|\pi,\rho,q_{s_i},Q_{EW}) w(\delta) \quad (4.1)$$

where $P(\delta|\pi,\rho,q_{s_i},Q_{EW})$ and $w(\delta)$ are the probability of receiving δ and the reward in that case, respectively, given that strategy $(\pi,\rho,q_{s_i},Q_{EW})$ is used. The above optimization is done under N channel constraints, namely $\sum_i r_{ij} \leq R_j$, D source coding constraints: $\sum_j r_{ij} \leq B/L$, and N power constraints $\rho_k \leq P_k$, where P_k is power constraint of the k-th transmitter [94] © 2013 IEEE.

Note that the optimization in (4.1) includes source optimization (q_{s_i}), relay power allocation (ρ), relay selection (π), and scheduling/channel coding (Q_{EW}). The above problem is a hard combinatorial problem that cannot be solved in real time. Thus, the following simplifications are introduced. First, assuming that the source block is pre-encoded, all q_{s_i}'s are fixed, and ρ, π, Q_{EW} are considered one at a time [94] © 2013 IEEE.

Power allocation ρ is done for each relay by looking only at the single relay-destination channel and selecting the lowest power level that will provide average packet loss rate below a threshold. If that is not possible, then $\rho_k = P_k$ is set. This strategy is justified by the fact that transmitter power is handled at the physical layer, to minimize the bit error rate for each link, and without any knowledge of the source [94] © 2013 IEEE.

Next, optimal π is found, assuming that no scheduling is done, i.e., all encoded packets of a description are treated equally. Then description *i* is decoded if at least $s_i \geq q_{s_i} B/L$ packets from this description are received. Note that for maximum separable codes, $s_i = \lceil q_{s_i} B/L \rceil$, with high probability, while for RLC codes s_i very slightly exceeds $\lceil q_{s_i} B/L \rceil$ (the number of source packets to be sent). In the analysis it is further assumed that R_j's and p_j's are constant during the transmission of one encoded source block. Even with fixed q_{s_i} and ρ, optimizing π is a difficult optimization problem. It is simplified by assuming that if the destination decodes description *i*, it gets reward w_i, where w_i is estimated based on the source recovery using only description *i*. The main motivation for this simplification is that it will reward descriptions that contribute highly to the reconstruction while significantly reducing optimization complexity. The total reward is the sum of the rewards for all decoded descriptions. Thus (4.1) is optimized restricting δ to be of cardinality 1. With a slight change of notation, for clarity (5.1) is written as:

$$\max_{\pi} \quad w(\pi) = \max_{\pi} \quad \sum_{i=1}^{D} P_i(\pi) w_i \tag{4.2}$$

where $P_i(\pi)$ and w_i are the probability that description *i* is recovered if π is used and the reward in that case, respectively. Next the algorithm is proposed that efficiently solves optimization problem in (4.2) [94] © 2013 IEEE.

4.5.1 Global Optimum – Algorithm 1

Without loss of generality, it is assumed that all available resources are allocated. Therefore,

$\sum_{i=1}^{D} r_{ij} = Rj$. Then, the resource allocation problem is

$RA(R_1,...,R_N, p_1,...,p_N ; s_1,...,s_D , w_1,...,w_D)$
whose optimal solution π^* is the one that maximizes (4.2) [94] © 2013 IEEE.

Brute force complexity. For a channel with Rj, there are $(R_j + 1)^D/D!$ ways of allocating the D descriptions. Therefore, the brute force complexity is $\sim \left(\prod_{j=1}^{N}(R_j + 1)\right)^D \sim O(R^{ND})$, where R is the geometric mean of Rj , j = 1,...,N [94] © 2013 IEEE.

Proposed dynamic programming solution. Note that if π^* is the optimal solution of

$RA(R_1,...,R_N, p_1,...,p_N ; s_1,...,s_D , w_1,...,w_D)$,

then the sub-solution defined by sub-matrix π^* (1 : D -1, :) has to be the optimal solution of

$$RA(R_1 - r_{D,1,...,} R_N - r_{D,N}, p_1, \ldots, p_N; s_{1,...} s_{D-1}, w_1, \ldots, w_{D-1}).$$

The advantage can be taken of this fact to decrease optimization complexity by dynamic programming. In a nutshell, the algorithm will first seek the optimal solutions for all combinations of $1 \leq R_j \leq R_j$ and all $1 \leq d \leq D$. At d = 1, the problem is trivial as the optimal allocations for $RA(R_1, \ldots, R_N, p_1, \ldots, p_N; s_1, w_1)$ is simply $\pi^* = [R_1, R_2, \ldots, R_N]$(i.e., allocate all resources to the first description) [94] © 2013 IEEE.

Assume now that solutions for d are available. For d = d + 1, computing optimal solutions of $RA(R_1, \ldots, R_N, p_1, \ldots, p_N; s_1, \ldots s_{d+1}, w_{1,\ldots,} w_{d+1})$ will only involve scanning through the $\Pi_i R_i$ solutions of the previous level. Therefore finding all optimal solutions for this level (d + 1), involves approximately $\left(\Pi_i(R_i + 1)\right)^2 /2$ computations [94] © 2013 IEEE.

Note that, to compute the optimal solution will require going through all D levels and thus have complexity $(D-1)\left(\Pi_j(R_j + 1)\right)^2 /2$ or approximately $D\left(\Pi_{j=1}^N(R_j + 1)\right)^2 \sim O(DR^{2N})$ instead of $O(DR^{ND})$, needed for the brute force method [94] © 2013 IEEE.

4.5.2 Fast Solution: Algorithm 2

Algorithm 1 finds the optimal solution to the allocation problem in (4.2). However, its complexity and memory requirements might still be too large for real-time applications, especially when N or R_j's are large. That is why; a simple and fast ad hoc strategy is also proposed. It will be shown in Section 4.6 that this strategy works well for the MDC video, where the descriptions do not vary largely in importance. First, the channels are arranged in the decreasing order of quality, i.e., such that $p_1 \leq p_2 \leq \cdots \leq p_N$ [94] © 2013 IEEE.

That is, Channel 1 (Ch1) is the best channel and so forth. Next, the descriptions are arranged such that $w_1 \geq w_2 \geq \cdots \geq w_N$. That is, D1 is the most important description, and so forth [94] © 2013 IEEE.

Without loss of generality it is assumed that D ≤ N. If D > N the descriptions could be merged to end up with total of N [94] © 2013 IEEE.

Algorithm 2 proceeds as follows: Put D_i into Ch_i either until (1) Ch_i is filled; or (2) $\frac{(1+\epsilon)s_i}{1-p_i}$ packets have been put, where ϵ is heuristically set to 10%. In the next step, channels 1,...,D that have not been filled are filled equally with the descriptions for which condition (1) occurred, or if there are no such descriptions, Ch_i is filled with description *i*. Channels D +1,...,N are filled equally with all descriptions [94] © 2013 IEEE.

The algorithm can be made blind of the source, by arranging descriptions in arbitrary order. Note that, Algorithm 2 is applicable to any channel model by properly selecting ϵ [94] © 2013 IEEE.

4.5.3 AL-FEC and Packet Scheduling

AL-FEC is a flexible software-based solution that can effectively combat fading and easily be tied to source coding via source-channel coding algorithms (see [55] and references therein). Although Digital Fountain Raptor codes [9], are one of the most efficient rateless packet loss codes, they equally protect the entire stream and have a sharp avalanche effect. Thus, they are not very suitable to applications where importance of the sent packets significantly varies, as is the case in video coding. EWF codes are a recent solution to this problem proposed in [21], [94] © 2013 IEEE.

RLC codes provide similar performance as Raptor codes for point-to-point communications [55] at the expense of increased decoding complexity. However, it is shown in [55] that if the used source block is small, RLC performs similarly to Raptor codes, and have acceptable decoding complexity via progressive GE [20] [36], [94] © 2013 IEEE.

The main advantage of RLC compared to Raptor codes manifests itself in multipath scenarios, as RLC effectively apply network coding concepts in the multi-node settings [13], [94] © 2013 IEEE.

The encoding process for EW-RLC starts by selecting a window from which the RLC encoded symbol is to be generated. These probabilities of window selections are organized into a vector Q_{EW} and are equivalent to channel coding rates for each source layer/window. After a window is selected, the encoding is the standard RLC encoding performed over the source packets contained in that particular window only [22]. In NOW scheme, the windows are non-overlapping and each window is protected with a separate AL-FEC code. The performance of EW-RLC schemes is concluded to be better than the corresponding NOW-RLC schemes, and this gain comes from the increased coding flexibility introduced by overlapping the windows [94] © 2013 IEEE.

In this section EW-RLC over l windows are employed from consecutive source blocks that are of unequal importance, i.e., each description is organized into l layers of unequal importance, which form l windows of the EW-RLC scheme [94] © 2013 IEEE.

To provide optimal allocation of redundancy to different source priority classes, for each description i one can maximize the expected reward using analytically computed probabilities of decoding error performance. That is,

$$\max_{Q_{EW}^{i}} w_i\left(Q_{EW}^{i}\right) = \sum_{k=0}^{l} P_i\left(k|Q_{EW}^{i}\right)w_i(k) \tag{4.3}$$

where Q_{EW}^{i} is an l-length vector of window selection probabilities that determines the UEP allocation scheme, $P_i\left(k|Q_{EW}^{i}\right)$ is the probability that layer k of description i will be the highest layer recovered if Q_{EW}^{i} is used, $P_i(0)$ is the probability that nothing is recovered, $w_i(k)$ is reward if all layers up to and including layer k are recovered, k = 1, ..., l. Analytical expressions for probabilities that a layer is recovered assuming a random channel loss model can be found in [22], [94] © 2013 IEEE.

4.6 Simulation Setup

In this section, the specific source-channel-network scheme is described along with the parameters used.

4.6.1 Video Coding and AL-FEC

Error resilience via slicing splits the frame into multiple slices that are independently encoded and separated by resynchronization points. Since slices can be made independently decodable, the partitioning of a frame into slices can be used to create multiple descriptions with fine granularity [89] © 2012 IEEE. For example, D = 2 descriptions, namely, MDC1 and MDC2, can be created by assigning three slices from each frame to each description. In this way each description has a significant representation of every frame which helps the decoding process. The way the slices are chosen for inclusion in a description is shown in Figure 4-7 [94] © 2013 IEEE.

Firstly, both descriptions have the intra-coded IDR [2] as the first entry. Thereafter, starting from the first following frame, alternate slices are copied to each description. This way, the overlap in the source content between two descriptions is minimal [94] © 2013 IEEE.

Table 4-11: Compression results (size in bytes and PSNR in dB) for the first GOP of the Foreman sequence [94] © 2013 IEEE.

Layer	MDC1			MDC2		
	Size	PSNR	Packets	Size	PSNR	Packets
BL	27,960	25.76	39	31,871	25.58	44
EL	23,980	26.79	33	23,539	26.2	33
Total	51,940	26.79	72	55,410	26.2	72

After video encoding, two independently decodable descriptions of possibly different sizes are generated. Each description contains $l = 2$ quality layers obtained by grouping IDR + DP A + DP B into one layer (i.e., the base layer (BL)) and DP C into the second quality enhancement layer (EL) [94] © 2012 IEEE.

Table 4-11 shows the peak signal-to-noise ratio (PSNR) and the rate (in bytes) for the first GOP of size 16 frames for the standard CIF Foreman video sequence. The video is encoded at 25 frames per sec and the total compression rate is 1.2 Mbps [94] © 2013 IEEE.

The best video quality of 41.07 dB is achieved when both MDCs are received completely. However, it can be seen from Table 4-11 that due to the way the descriptions have been designed, the video can be decoded with the BL of any one description only. This layering of video data within a description makes this scheme adaptive and practical for varying channel conditions. In the case where both descriptions are lost entirely and nothing is decodable for the GOP then the decoder applies a simple error concealment technique by replacing the lost GOP by the last frame of the previously successfully decoded GOP. Each description is independently protected against packet losses by an EW-RLC. The BL forms the first window, W1, and the (BL+EL) forms the second window, W2 [94] © 2013 IEEE.

4.6.2 Simulation Parameters

The BS compresses the video, performs RLC protection, relay selection based on the feedback from the relay regarding the quality of the relay-destination link and the relay transmit power, and schedules encoded packets for transmission [94] © 2013 IEEE.

The setup with two relays and a direct link from the BS to the destination is considered, thus N = 3. To model the fading channel with two femto-relays, FSMC is used that reflects the change in the bit error rate depending on Markov chain state-space resulting in different received data rates. In the model each wireless Rayleigh block-fading channel is represented by a packet-level FSMC channel model [81] that adapts with the mobility of the user and transmit power [94] © 2013 IEEE.

Symbol sizes of 734 bytes are used for RLC. A header overhead of 60 bytes per packet is assumed to cater for the headers added at the various protocol layers, e.g., RTP/UDP/IP (note however that robust header compression can significantly reduce the size of the header). Only one symbol is placed in the packet, thus transmission packets are of size 734+60 = 794 bytes. GE decoding is performed on source block sizes of 70-80. These small block sizes enable fast decoding on a smart-phone without sacrificing performance [20]. The reconstructed quality is assessed using PSNR and optimization (Algorithms 1 and 2) is performed with reconstructed frame-average PSNR as reward w [94] © 2013 IEEE.

4.6.3 Simulation Results and Discussion

First, we look at the system with N = 2 relays. The distances between the destination and BS, femto relay 1 and femto relay 2 are 70m, 25m, and 20m, respectively [75]. The data transmission rate is 1000 kbps, 600 kbps and 550 kbps for the BS, femto 1 and femto 2 relays, respectively. This corresponds to the maximum number of encoded packets that these paths can take during one source block (one encoded GOP) of 100, 60, and 55, respectively [94] © 2013 IEEE.

The simulations have been done for two different power constraints for the two femto relays: in Configuration 1 (C1) the required average packet loss rates are set in the femto relay 1 and 2 links to 0.25 and 0.3, respectively. This results in physical transmitter powers of 30 dBm and 35 dBm, respectively, which translates into the averaged received signal-to-noise ratio (SNR) of 23 dB and 33 dB, respectively. In Configuration 2 (C2) the required average packet loss rates are set in the femto relay 1 and 2 links to be 0.05 and 0.1, respectively. This results in the physical transmitter power of 47 dBm, for both relays, which translates into the average received SNR of 40 dB and 44 dB, respectively [94] © 2013 IEEE.

For the above femto relay transmit power constraints, the BS power level is varied in the range of 40 to 70 dBm, which results in the average received SNR between 12 dB and 43 dB. Then, the average packet loss rate in the direct link varies from 0.05 to 0.3. Several UEP EW-RLC schemes are selected, obtained by changing Q_{EW} to test their robustness to varying channel conditions. Note that the same Q_{EW} is always used for the two descriptions [94] © 2013 IEEE.

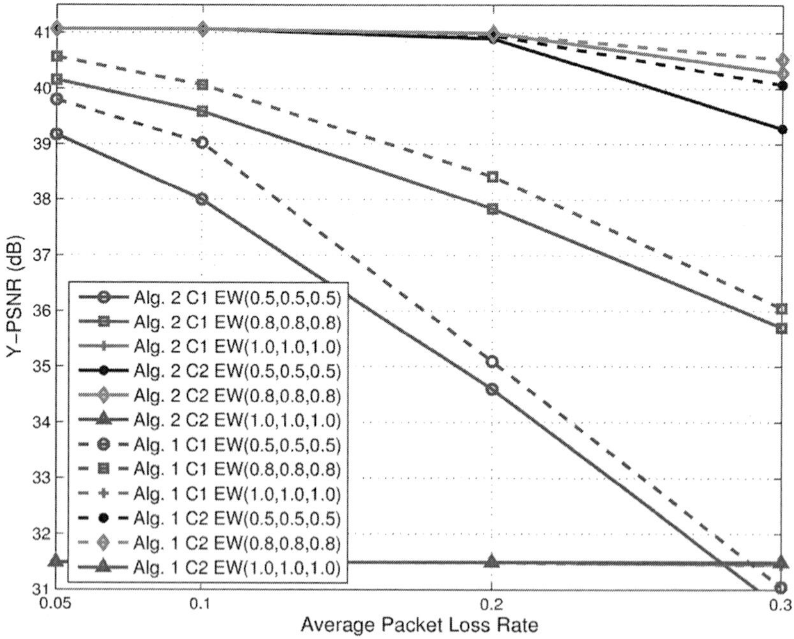

Figure 4-12: Frame average Y-PSNR vs. average packet loss rate for Configuration (1) and (2) for the FSMC channel – Foreman sequence. [94] © 2013 IEEE.

The results for a mobile speed of 3km/hr, comparing Algorithms 1 and 2 are shown in Figure 4-12. The term EW(x, y, z) refers to a set of selection probabilities for windows W1 (BL), such that the selection probability of the BL for both descriptions for the direct link, link via Relay 1 and link via Relay 2 are x, y, and z, respectively. It can be seen from the figure that the two algorithms show similar performance. Indeed, Algorithm 1 only performs 0.5-1 dB better. Note that, in the EW(1.0,1.0,1.0) scheme only window W1 is selected, hence the PSNR cannot go beyond that achievable with the two BLs [94] © 2013 IEEE.

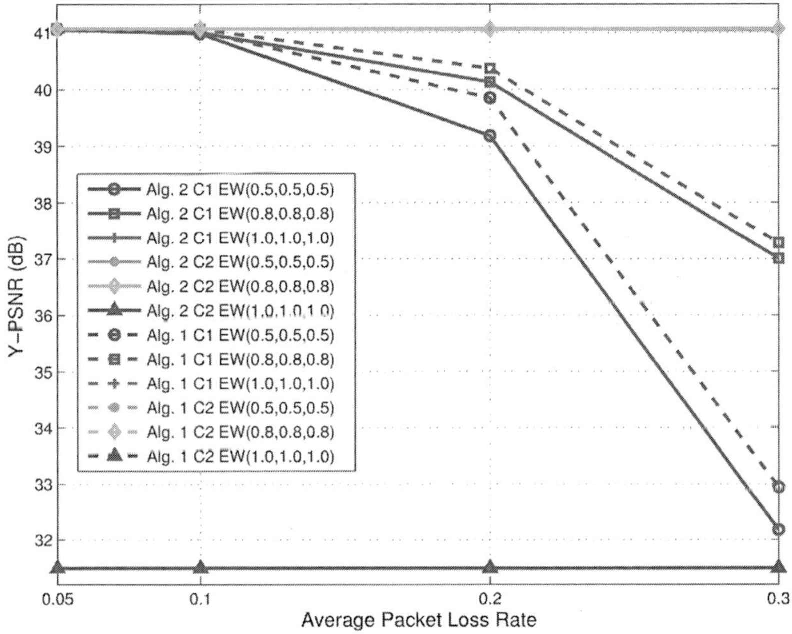

Figure 4-13: Frame average Y-PSNR vs. average packet loss rate for Configuration (1) and (2) for the random loss channel model- Foreman sequence. [94] © 2013 IEEE.

Setting the probability of selection of the BL to 0.8 (Q_{EW} = (0.8, 0.2)) provides the best results for all source power levels. Note that in this case, 80% of rate is allocated to the BL, but even with a loss rate as high as 30%, ELs are often recovered showing a PSNR gain of 4 dB compared to the case when only BL is sent alone. For Configuration 2, setting the probability of selection of the BL to 0.5 or 0.8 provides the best results. Algorithm 1 provides a negligible performance gain over Algorithm 2 [94] © 2013 IEEE.

The results for the random loss channel model are shown in Figure 4-13. Note that, the same average channel loss rates are used as those given by the FSMC model. As expected, the performance of Algorithm 1 is consistently better than that of Algorithm 2. However, the difference is very small (under 1dB). The EW(1.0,1.0,1.0) scheme again performs poorly, since it does not exploit the available bandwidth to the fullest. Indeed, a gain of 7-10 dB can be observed by sending ELs. The results for the random channel are only up to 0.5dB better than those for the FSMC channel, which demonstrates that the proposed schemes are robust to the bursty nature of the FSMC channel. For Configuration 2, where the loss rates over the relay channels are much lower, both algorithms perform equally well [94] © 2013 IEEE.

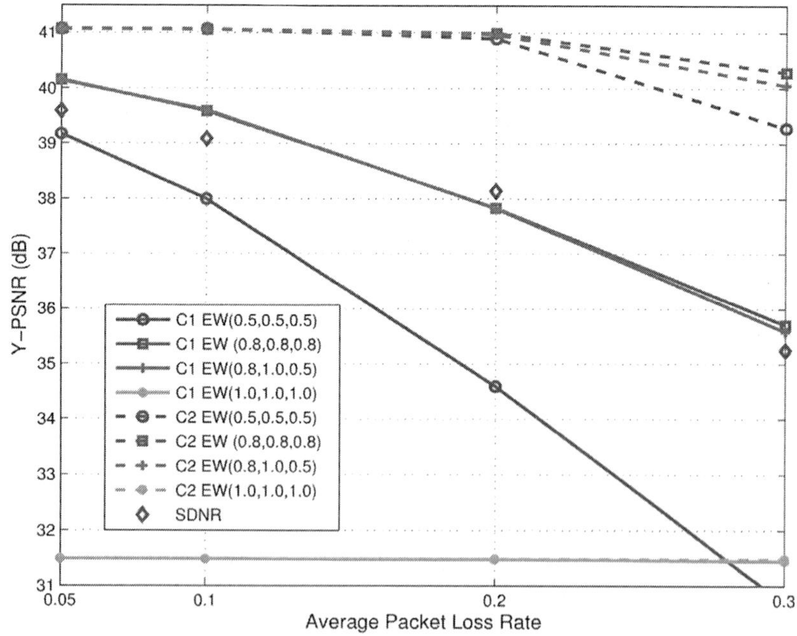

Figure 4-14: Frame average Y-PSNR vs. average packet loss rate for both configurations for the proposed Algorithm 2 and the single description no relay scheme for the FSMC channel – Foreman sequence. [94] © 2013 IEEE.

In the next two figures results of Algorithm 2 are compared to the benchmark scheme - an adaptive source-optimized single description no relay (SDNR) scheme that transmits a single-layer AVC (without MDC) For each testing channel condition we find the optimal source rate (via exhaustive search over all video source rates between 500kbps and 1.5Mbps where the lower and upper video source rates were found to provide too little and too much protection, respectively, and were never selected); thus this scheme adaptively changes the source rate to match the channel conditions. It is assumed that perfect knowledge of the channel is available prior to source coding, which is not the case for the proposed schemes [94] © 2013 IEEE.

Figure 4-14 shows results for the FSMC channel and both configurations and four different scheduling strategies. It can be seen that Algorithm 2 significantly outperforms the single layer no relay schemes if the packet scheduling is properly done. Note that the benchmark schemes do not use MDC, and hence provide better performance in the error-free scenarios. Using relaying, power is effectively distributed among three nodes providing increased performance compared to the SDNR schemes. It can also be observed that by increasing the power levels of the two relays for 9 dB in total, a gain of roughly 5 dB at the highest channel loss rate and 0.5 dB for the lowest channel loss rate is obtained [94] © 2013 IEEE.

The results for the random loss channel model are shown in Figure 4-15. The optimized relay-assisted scheme outperforms by almost 2 dB the no-relay scheme at the lowest packet loss rate and by almost 5 dB for the highest packet loss rate. Results for Configuration 2 are up to 5 dB better than those for Configuration 1. For both type of channels, Configuration 1 scheme at a packet loss rate of 0.3 shows roughly the same performance as the source-optimized benchmark scheme, because in this case, there is only little use of relaying due to high loss rates at the two relay links [94] © 2013 IEEE.

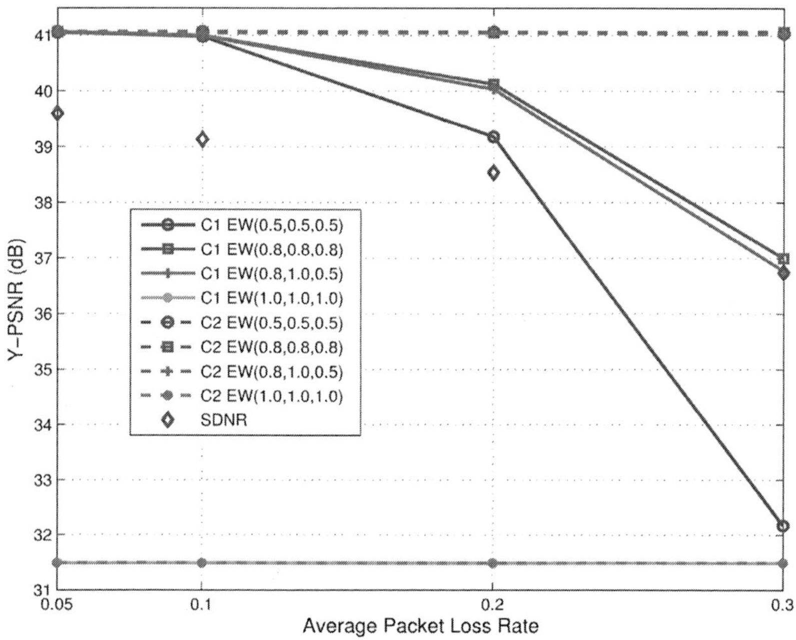

Figure 4-15: Frame average Y-PSNR vs. average packet loss rate for both configurations for the proposed Algorithm 2 and the single description no relay schemes for the random loss channel model [94] © 2013 IEEE.

The above simulations clearly indicate that the relay-assisted schemes are very robust to a change in channel conditions of the direct and relay links. The sub-optimal and fast Algorithm 2 shows close performance to Algorithm 1. It is important to send both BL and EL for both descriptions to obtain high performance. Scheduling packets at the source (varying Q_{EW}) is important at the high packet loss rates and can provide a gain of over 10 dB [94] © 2013 IEEE.

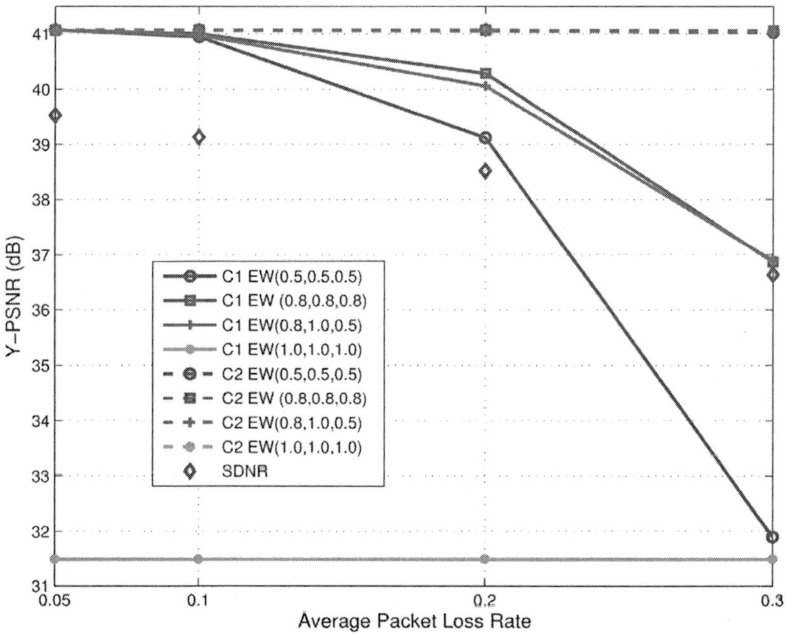

Figure 4-16: Frame average Y-PSNR vs. average packet loss rate for both configurations for the proposed Algorithm 2 and the single layer schemes for the FSMC channel for the mobile speed of 30 km/hr [94] © 2013 IEEE.

Figure 4-16 shows results for the FSMC channel and the mobile speed of 30km/hr. Comparing Figures 4-14 and 4-16, one can see that the proposed solutions are very robust to an increase in mobile speed [94] © 2013 IEEE.

(a) M1M2, PSNR =41.56dB.

(b) H1H2, PSNR =35.69dB.

(c) M1H2, PSNR =38.22dB.

(d) H1M2, PSNR =37.68dB.

(e) M1, PSNR = 33.77dB.

(f) M2, PSNR = 32.9dB.

(g) H1, PSNR =32.9dB.

(h) H1, PSNR =32.41dB.

Figure 4-17: Frame 13 for Container sequence for different combinations.

Figure 4-17 shows frame 13 for the Container sequence. The two-letter caption such as M1H2 implies that MDC1 has been received in full whereas for MDC2 only its HPC has been received. Similarly, one-letter caption such as H1 means that only HPC of MDC1 has been received, whereas MDC2 has been completely lost. Each of these 8 combinations is sufficient in itself to re-create the whole 16 frames of the original GOP.. It is interesting to note that despite no perceptible visual difference between the images, the PSNR values are significantly different.

4.7 Conclusion

The real-time transmission of layered MDC video over relay-assisted paths is considered. MDC scheme is designed using slicing and data partitioning features of H.264/AVC and the resulting packets are fed into the EW-RLC encoder for erasure protection. The encoded packets were streamed over direct link and over two relay-assisted channels. A resource optimization framework was developed, using dynamic programming and the proposed fast algorithm that sub-optimally select relays and schedule packets for transmission [94] © 2013 IEEE.

There could be an impact on battery life for multi-interface solution however it is small as compared to video decoding and display itself. Also, the solution proposed here is based on EW RLC which is comparable to Raptor codes in complexity. Raptor codes are already employed in phones so a short code length EW RLC should not have a large impact on such solutions.

CHAPTER FIVE

FUTURE OF VIDEO STREAMING

5.1 Introduction

The field of video streaming has seen tremendous growth. With the deployment of high capacity communication networks such as 5G, optical networks and DVB-NGH it is likely that novel applications would follow. Hence there would always be a need for efficient solutions which address the challenges of multimedia communication over the emerging wireless networks. With the confluence of cellular and broadcast communication, fountain codes, such as, RLC would be increasingly important.

Application Layer FEC has been extensively used to provide reliable video streaming over wireless channels. Rateless codes can prove to be advantageous in such scenarios as potentially unlimited number of encoded symbols can be generated, from a given set of source symbols. Such codes can provide additional benefit if the unequal importance within encoded video data is taken into consideration. The DP and slicing feature of H.264/AVC video have been extensively used in this text to design schemes for providing unequal protection to different video data elements.

Any improvement in network bandwidth and processing power due to deployment of new technologies gets quickly consumed by newer and innovative video streaming applications that then become possible. This interaction between better video support and increased expectations of the users drives the need to design better and innovative solutions for error resilience, error correction and for exploiting multiple communication interfaces available on the latest mobile hand held devices.

IP pervasiveness in devices and the accompanied technological advances are opening newer application domains for video streaming. This chapter covers few of the current innovations from video streaming perspective.

5.2 Emerging Support for Video Streaming

5.2.1 Dynamic Adaptive Streaming over HTTP (DASH)

Video streaming over UDP protocol has been the favoured approach however it changed with DASH, where the streaming takes place over TCP. The basic idea is to pre-encode the video into different data rates and resolutions and make them available at the server. The client uses HTTP GET requests to the server for the appropriate video segments. The requests initially take place in succession till such time the client buffer gets filled. Thereafter more segments get requested only once there is space created in the buffer due to play out.

5.2.2 Mobile devices

The mobile devices enable streaming live video. With better network support and better devices this trend will continue to drive the field forward. The devices now have multiple interfaces and provide many services and a rich feature set. Mobile video streaming is poised to become the dominant video traffic.

5.2.3 3D Video Delivery

The public demand to generate and consume better and better video has resulted in the 3D video delivery in television sets, cinemas and mobile phones. 3D video streaming is expected to be more popular in the future as 3D capture/display devices become common.

5.2.4 Cloud based Services

Cloud services are already re-shaping how the content is generated and distributed. The movement of video streaming to cloud based services has many advantages, the major being that there is no upfront hardware investment required for setting up the servers and network. The cloud based services also provide built-in fault tolerance, backup and recovery mechanism.

5.2.5 webRTC

The webRTC has the promise of integrating the delivery of video from within the web browser without the need of any plug-ins. Thus the video delivery would be seamless and video streaming will be possible from within the web page.

5.2.6 RTP Media Congestion Avoidance Techniques (RMCAT)

With the introduction of WebRTC the trend in supporting video over UDP is likely to continue. In such cases there is a need to standardise the congestion control for RTP applications which are substantially different from the congestion control algorithms of the TCP. The finalisation of such techniques will aid in further deployments of video streaming.

5.3 Conclusion

This chapter highlights the emerging technologies which support the growing field of video streaming. There has been tremendous growth in network technologies and video compression algorithms but on the other hand better user experience results in further demands for innovative applications. Therefore, better video compression algorithms, error-resilience techniques, and error correction coding schemes will be required to cope with the insatiable need of the video streaming user community.

The newer applications need more bandwidth and also because of widespread use of wireless mobile devices the bandwidth is not likely to cope with the applications' requirements. Hence the techniques such as those described in this book for protection of video data against packet losses to meet the strict time and delay constraints of video data would always be important.

BIBLIOGRAPHY

[1] "Video Streaming: Concepts, Algorithms, and Systems", J.G. Apostolopoulos, W. Tan, and S.J. Wee, *Handbook of Video Databases*, CRC Press, March 2003.

[2] T. Wiegand, G. J. Sullivan, G Bjøntegaard, and A. Luthra, "Overview of the H.264/AVC Video Coding Standard", *IEEE Trans. on Circuits and Systems for Video Technology,* vol. 13, no. 7, pp. 560-576, Jul. 2003.

[3] T. Wiegand, H. Schwarz, A. Joch, F. Kossentini, and G. J. Sullivan, "Rate-constrained coder control and comparison of video coding standards," *IEEE Trans. Circuits Syst. Video Technol.*, vol. 13, no. 7, pp. 688–703, Jul. 2003.

[4] H. Schwarz, D. Marpe, and T. Wiegand, "Overview of the Scalable Video Coding Extension of the H.264/AVC Standard," *IEEE Trans. on Circuits and Systems for Video Technology*, vol. 17, no. 9, pp. 1103-1120, Sept 2007.

[5] S. Wenger, "H.264/AVC over IP," *IEEE Trans. Cir. Syst. Video Technol.*, vol. 13, pp. 645–656, July 2003.

[6] M. van der Schaar, *Multimedia Over IP and Wireless Networks*, Academic Press, 2007.

[7] M. Mitzenmacher, "Digital Fountains: A Survey and Look Forward," *IEEE Information Theory Workshop*, pp. 271-276, 2004.

[8] M. Luby, "LT codes", *Proc. 43rd Annual IEEE Symposium on Foundations of Computer Science,* pp. 271-280, 2002.

[9] A. Shokrollahi, "Raptor codes," *IEEE Trans. Inform. Theory,* vol. 52, no. 6, pp. 2251-2567, 2006.

[10] D. J. C. Mackay, "Fountain Codes," *IEE Proceedings on Communications*, vol. 152, pp. 1062-1068, 2005.

[11] D. Gomez-Barquero, D. Gozalvez, and N. Cardona, "Application layer FEC for mobile TV delivery in IP datacast over DVB-H systems," *IEEE Trans. Broadcasting*, vol. 55, no. 6, pp. 396-406, 2009.

[12] "ETSI TS 126 346, universal mobile telecommunications system (umts); multimedia broadcast/multicast service (mbms); protocols and codecs," *ETSI Tech. Spec.*, 2005.

[13] D. Lun, M. Medard, R. Koetter, and M. Effros, "On coding for reliable communication over packet networks," *Physical Communication*, vol. 1, no. 1, pp. 22–30, 2008.

[14] D. Vukobratović and V. Stanković, "Unequal Error Protection Random Linear Coding Strategies for Erasure Channels", *IEEE Trans. on Comm,* vol. 6, no. 5, pp 1243-1252, 2012.

[15] Y. Lin, B. Li, and B. Liang, "Priority random linear codes in distributed storage systems," *IEEE Trans. Parallel Distributed Syst.,* vol. 20, no. 11, pp. 1653–1667, Nov. 2009.

[16] K. Nguyen, T. Nguyen, and S. Cheung, "Video streaming with network coding," *Springer - J. Signal Process. Syst.,* vol. 59, no. 3, pp. 319–333, 2010.

[17] M. Halloush and H. Radha, "Network coding with multigeneration mixing: Analysis and applications for video communication," *Proc. IEEE ICC 2008,* Beijing, China, May 2008.

[18] X. Liu, G. Cheung, and C-N. Chuah, "Structured network coding and cooperative wireless ad-hoc peer-to-peer repair for WWAN video broadcast," *IEEE Trans. Multimedia,* vol. 11, no. 4, pp. 730–741, 2009.

[19] N. Thomos, J. Chakareski and P. Frossard, "Randomized network coding for UEP video delivery in overlay networks," *Proc. IEEE ICME 2009,* New York, NY, USA, July 2009.

[20] H. Shojania and B. Li, "Random network coding on the iphone: Fact or fiction?," in *NOSSDAV 2009, Williamsburg, USA*, June 2009.

[21] D. Vukobratović, V. Stanković, D. Sejdinović, L. Stanković, and Z. Xiong, "Expanding window fountain codes for scalable video multicast," *IEEE Trans. Multimedia*, vol. 11, no. 6, pp. 1094–1104, 2009.

[22] D. Vukobratović and V. Stanković, "Unequal error protection random linear coding for multimedia communications," in *Proc. MMSP-2010 IEEE Multimedia and Signal Processing Workshop,* Saint Malo, France, October 2010.

[23] G. Faria, J. A. Henriksson, E. Stare, and P. Talmola, "DVB-H: Digital broadcast services to handheld devices," *Proc. of the IEEE,* vol. 94, no. 1, pp. 194-209, Jan. 2006.

[24] DVB-CM-NGH015R1 Version 1.01, "Commercial Requirements for DVB-NGH", 15 June 2009.

[25] L. Vangelista et al., "Key technologies for next-generation terrestrial digital television standard DVB-T2", *IEEE Commun. Magazine,* vol 47. no. 10, pp. 146-153, Oct. 2009.

[26] Y.Yang, H. Hiu, J. Xu, and G. Mao, "Relay technologies for WIMAX and LTE-A mobile system," *IEEE Commun. Magazine*, Oct. 2009.

[27] 3GPP TR 36.814 V1.2.1, "Further Advancements for EUTRA: Physical Layer Aspects," Tech. Spec. Group Radio Access Network Rel. 9, June 2009.

[28] TM-NGHSM055r1 TM-H NGH Study Mission Report.

[29] E. Dahlman, S. Parkvall, and J. Skold, "*4G LTE/LTE-Advanced for mobile broadband*" Academic Press, 2011.

[30] V.K. Goyal, "Multiple description coding: Compression meets the network," *IEEE Signal Processing Magazine*, vol. 18, pp. 74-93, 2001.

[31] ETSI EN 302 755, "Digital Video Broadcasting (DVB); Frame structure channel coding and modulation for a second generation digital terrestrial television broadcasting system (DVB-T2)", *V1.3.1, DRAFT,* Oct. 2011.

[32] T. Stockhammer, A. Shokrollahi, M. Watson, M. Luby, and T. Gasiba, *Application Layer FEC for Mobile Multimedia Broadcasting (in Handbook of Mobile Broadcasting),* CRC Press (Eds: B. Furht, and S. Ahson), 2008.

[33] M. Luby, T. Gasiba, T. Stockhammer, and M. Watson, "Reliable multimedia download delivery in cellular broadcast networks," *IEEE Trans. Broadcast.,* vol. 53, no. 1, pp. 235-246, Mar. 2007.

[34] C. Hellge, D.Gomez-Barquero, T. Schierl, and T.Wiegand, "Layer-aware forward error correction for mobile broadcast of layered media," *IEEE Trans. Multimedia,* vol. 13, pp. 551-562, Jun 2011.

[35] "ETSI TR 102 591, digital video broadcast (DVB); IP datacast: Content delivery protocols (CDP) implementation guidelines part 1: IP datacast over DVB," ETSI Tech. Spec., 2010.

[36] J. Jin, B. Li, and T. Kong, "Is random network coding helpful in wimax?," *in Proc. IEEE INFOCOM 2008,* Phoenix, AZ, April 2008.

[37] R. Ahlswede, N. Cai, S. R. Li, and R. W. Yeung, "Network information flow," *IEEE Trans. Inform. Theory,* vol. 46, no. 4, pp. 1204-1216, 2000.

[38] Q. Zhang, F. Fitzek, and M. Katz, "On the energy saving potential in DVB-H networks exploiting cooperation among mobile devices," *in Proc. Cognitive Wireless Networks,* 2007, pp. 473–484.

[39] L. Benacem and S.D. Blostein, "Raptor-network coding enabled strategies for energy saving in DVB-H multimedia

communications," *in Proc. 1st Int. Conf. Green Circuits and Systems,* pp. 527–532, June 2010.

[40] T. Schierl, T. Stockhammer, and T. Wiegand, "Mobile Video Transmission using Scalable Video Coding," *IEEE Trans. Circuits and Sys. Video Techn.,* vol. 17, pp. 1204-1217, Sept. 2007.

[41] A. T. Connie, P. Nasiopoulos, Y.P. Fallah, and V. C.M. Leung, "SCTP-based transmission of data-partitioned H.264 video," *In Proc. of the 4th ACM workshop on Wireless multimedia networking and performance modeling*, pp 32-36. ACM, 2008.

[42] T. Stockhammer, and M. Bystrom, "H.264/AVC data partitioning for mobile video communication," *ICIP-2004 IEEE International Conference on Image Processing*, Brussels, Belgium, September 2004.

[43] D. Gomez-Barquero, K. Nybom, D. Vukobratović, and V. Stanković, "Scalable video coding for mobile broadcasting DVB system," *in Proc. ICME-2010 IEEE Intl. Conf. Multimedia and Expo,* Singapore, July 2010.

[44] W. Chung, S. Paluri, S. Kumar, S. Nagaraj, "Unequal Error Protection for H.264 Video using RCPC Codes and Hierarchical QAM," *in Proc. ICC-2010 IEEE Intl. Conf. on Comm.*, 2010.

[45] B. Barmada, M.M. Ghandi, E.V. Jones, and M. Ghanbari, "Prioritized transmission of data partitioned H.264 video with hierarchical QAM," *IEEE Signal Processing Letters,* vol. 12. no. 8, pp. 577-580, Aug. 2005.

[46] L. Al-Jobouri, M. Fleury, and M. Ghanbari, "Error-resilient IPTV for an IEEE 802.16e channel," *Wireless Engineering and Technology,* pp. 70-79, April 2011.

[47] R. Razavi, M. Fleury, M. Altaf, H. Sammak, and M. Ghanbari, "H.264 video streaming with data-partitioning and Growth codes," *in Proc. ICIP-2009 Intl. Conf. Image Proc.,* Cairo, Egypt, Oct. 2009.

[48] A. Bouabdallah and J. Lacan, "Dependency-Aware Unequal Erasure Protection Codes," *J. Zheijiang Univ. Science A,* vol. 7, pp. 27-33, Apr. 2006.

[49] D. Sejdinović, D. Vukobratović, A. Doufexi, V. ˇSenk, and R. Piechocki, "Expanding Window Fountain codes for unequal error protection," IEEE Trans. Commun., vol. 57, pp. 2510-2516, Sept. 2009.

[50] D. Vukobratović and V. Stanković, "On unequal error protection random linear coding for scalable video broadcasting," *in Proc. Packet-Video Workshop-2010,* Hong Kong, Dec. 2010.

[51] Y. Dhondt, S. Mys, K. Vermeirsch and R. Van de Walle, "Constrained Inter Prediction: Removing Dependencies between Different Data Partitions," *Proceedings of Advanced Concepts for Intelligent Visual Systems*, Delft, 28-31 August 2007, pp. 720-731.

[52] K. Nybom, S. Gronroos, and J. Bjorkqvist, "Expanding window fountain coded scalable video in broadcasting," *in Proc. ICME-2010 IEEE Int. Conf. Multimedia and Expo,* Singapore, July 2010.

[53] J. Poikonen and J. Paavola, "Error models for the transport stream packet channel in the DVB-H link layer", *in Proc. ICC-2006 IEEE Int. Conf. Commun.*, pp. 1861–1866, Istanbul, Turkey, 2006.

[54] H.264/AVC Reference Software, http:/iphome.hhi.de.

[55] S. Nazir, D. Vukobratović, and V. Stanković, "Performance evaluation of raptor and random linear codes for H.264/AVC video transmission over DVB-H networks," *in Proc. ICASSP-2011 IEEE Intl. Conf. Acoustics Speech and Sig. Proc.,* Prague, Czech Republic, May 2011.

[56] P. Vingelmann, F.H.P. Fitzek, M.V. Pedersen, J. Heide and H. Charaf, "Synchronized multimedia streaming on the iPhone platform with network coding," *in Proc. CCNC IEEE Consumer Commun. and Networking Conf.,* Las Vegas, USA, Jan. 2011.

[57] H. Shojania and B. Li, "Parallelized progressive network coding with hardware acceleration," *in Proc. IWQoS-2007 the 15th Intl. Workshop Quality of Service,* Chicago, IL, July 2007.

[58] S. Nazir, V. Stanković and D. Vukobratović, "Unequal error protection for data partitioned H.264/AVC video streaming with Raptor and Random Linear Codes for DVB-H networks," *in Proc. ICME-2011 IEEE Intl. Conf. Multimedia and Expo,* Barcelona, Spain, July 2011.

[59] E. Baccaglini, T.Tillo, and G. Olmo, "Slice Sorting for Unequal Loss Protection of Video Streams," *IEEE Signal Proc. Letters*, vol. 15, pp. 581-584, June 2008.

[60] A. Albanese, J. Blomer, J. Edmonds, M. Luby, M. Sudan, "Priority encoding transmission," IEEE Trans. On Info. Theory, vol. 42, pp1737-1744, Nov 1996.

[61] R. Hamzaoui, V. Stanković, and Z. Xiong, "Optimized error protection of scalable image bitstreams," *IEEE Signal Proc. Magazine,* vol. 22, pp. 91-107, Nov. 2005.

[62] S. Ahmad, R. Hamzaoui, M. Al-Akaidi, "Unequal error protection using LT codes and block duplication," *Proc. MESM 08, Middle Eastern Multiconference on Simulation and Modelling*, Amman, Aug. 2008.

[63] A.G. Dimakis, J. Wang, and K. Ramchandran, "Unequal Growth codes: Intermediate performance and unequal error protection for video streaming", *in Proc. MMSP-2007 IEEE Multimedia Signal Proc.*, pp. 107-110, Oct. 2007.

[64] L. Al-Jobouri, M. Fleury, and M. Ghanbari, "Adaptive rateless coding for data-partitioned video streaming over a broadband wireless channel", *6th Conference on Wireless Advanced (WiAD),* pp. 1–6, June 2010.

[65] N. Rahnavard, B. N. Vellambi, and F. Fekri, "Rateless codes with unequal error protection property", *IEEE Trans. Inf. Theory*, vol. 53, no. 53, pp. 1521–1532, Apr. 2007.

[66] J. Goshi, A. E. Mohr, R.E. Ladner, E. A. Riskin, "Unequal Loss Protection for H.263 Compressed Video," in *Proc. DCC-2003 IEEE Data Compression Conference,* 2003.

[67] L. Al-Jobouri, M. Fleury, and M. Ghanbari, "Channel Coding with Packet Duplication for Video Streaming over a Broadband Wireless Link," wireless advanced 2011.

[68] S. Nazir, D. Vukobratović, and V. Stanković, "Scalable broadcasting of sliced H.264/AVC over DVB-H network," *in Proc. ICON-2011 IEEE Intl. Conference on Networks, Special Session on Robust and Scalable Multimedia Networking,* Singapore, Dec. 2011.

[69] J. G. Apostolopoulos, "Reliable video communication over lossy packet networks using multiple state encoding and path diversity," in *Proc. SPIE Proc. Visual Commun. and Image Proc.,* vol. 4310, no. 1, 2001, pp. 392-409.

[70] V. Stankovic, R. Hamzaoui, and Z. Xiong, "Robust layered multiple description coding of scalable media data for multicast," *IEEE Signal Processing Letters,* vol. 12, pp. 154-157, Feb. 2005.

[71] "LTE-Advanced," 3GPP Global Initiative, http://www.3gpp.prg/LTE-Advanced, 2010.

[72] S. Parkvall, A. Furuskar, and E. Dahlman, "Evolution of LTE towards IMT-advanced," *IEEE Commun. Magazine,* pp. 84-91, Feb. 2011.

[73] H. Luo, S. Ci, D. Wu, J. Wu, and H. Tang, "Quality-driven cross-layer optimized video delivery over LTE," *IEEE Commun. Magazine,* pp. 102-109, Feb. 2010.

[74] S. Mao, S. Lin, S. Panwar, Y. Wang, and E. Celebi, "Video transport over ad hoc networks: Multistream coding with multipath transport," *IEEE J. Sel. Areas in Communs.,* vol. 21, no. 10, pp. 1721-1737, 2003.

[75] A. Agrawal, "Heterogeneous networks. A new paradigm for increasing cellular capacity," http://netseminar.stanford.edu/seminars/01 29 09.pdf, Jan. 2009.

[76] A.F. Molisch, N.B. Mehtra, J.S. Yedidia, J. Zhang, "Cooperative relay networks using fountain codes," *in Proc. IEEE GLOBECOM 2006*, pp. 1-6, 2006.

[77] T. Tillo, M. Grangetto, and G. Olmo, "Redundant slice optimal allocation for H.264 multiple description coding," *IEEE Trans. Circuits and Sys. for Video Techn.,* vol. 18, no. 1, pp. 59-70, Jan. 2008.

[78] Y. Liao and J.D. Gibson, "Routing-aware multiple description coding with multipath transport for video delivered over mobile ad-hoc networks," in *Proc. IEEE GLOBECOM-2010,* Miami, Dec. 2010.

[79] L. Zhou, B. Geller, B. Zheng, A. Wei, and J. Cui, "System scheduling for multi-description video streaming over wireless multi-hop networks," *IEEE Trans. Broadcasting,* vol. 55, no. 4, pp. 731-741, Dec. 2009.

[80] HD Video LTE-A Berlin Testbed: http://www.hhi.fraunhofer.de/en/departments/wireless-communication-and-networks/ video / cooperative-systems-hd/.

[81] Q. Zhang and S. Kassam, "Finite-State Markov model for Rayleigh fading channels," *IEEE Trans.Commun*, vol. 47, no. 11, pp. 1688-1692, Nov. 1999.

[82] K.T. Phan, L.-N. Tho, S.A. Vorobyov, and C. Tellambura, "Power allocation in wireless multi-user relay networks," *IEEE Trans. Wireless Commun.,* vol. 8, pp. 2535-2545, May 2009.

[83] K. Bakanoglu, S. Tomasin, and E. Erkip, "Resource allocation in wireless network with multiple relays," in *Proc. Asilomar 2008*, Pacific Grove, CA, 2008.

[84] E. Calvo, J. Vidal, and J.R. Fonollosa, "Optimal resource allocation in relay-assisted cellular networks with partial CSI," *IEEE Trans. Signal Proc.,* vol. 57, pp. 2809-2823, July 2009.

[85] D. Jurca and P. Frossard, "Video packet selection and scheduling for multipath streaming," *IEEE Trans. Multimedia,* vol. 9, pp. 629-641, April 2007.

[86] J.H. Sorensen, J. Astergaard, P. Popovski, and J. Chakareski, "Multiple description coding with feedback based network compression," in *Proc. IEEE GLOBECOM-2010,* Miami, FL, Dec. 2010.

[87] Y. Liu, "A Low Complexity Protocol for Relay Channels Employing Rateless Codes and Acknowledgement," *IEEE Intl Symposium on Info Theory*, pp. 1244-1248, 2006.

[88] J. Li, Z. Hu, Y. Wang, "DF/AF cooperative relay in LTE-A," *IEEE Vehicular Tech. Conf.*, pp. 1-5, 2010.

[89] S. Nazir, V. Stankovic, and D. Vukobratovic, "Adaptive Layered Multiple Description Coding for Wireless Video with Expanding Window Random Linear Codes," *ICASSP-2012 IEEE International Conference on Acoustic, Speech, and Signal Processing,* Kyoto, Japan, March 2012.

[90] S. Nazir, D. Vukobratovic, and V. Stankovic, "Expanding Window Random Linear Codes for Data Partitioned H.264 Video Transmission over DVB-H Network," *ICIP-2011 IEEE International Conference on Image Processing*, Brussels, Belgium, September 2011.

[91] S. Nazir, D. Vukobratovic, V. Stankovic, I. Andonovic, K. Nybom, and S. Gronroos, "Unequal Error Protection for Data-partitioned H.264/AVC Video Broadcasting," *Springer Multimedia Tools and Applications*, 2014.

[92] S. Nazir, D. Vukobratović, V. Stanković, and I. Andonović, "Rate Adaptive Selective Segment Assignment for Reliable Wireless Video Transmission," *International Journal of Digital Multimedia Broadcasting, vol. 2012*, Article ID 160521, 9 pages, 2012. doi:10.1155/2012/160521

[93] S. Nazir, F. Verdicchio, and G. Fairhurst, "Scalable Video Streaming Techniques over Cooperative Relay Networks", *International Journal of Computer Applications*, Vol.66 2013.

[94] S. Nazir, V. Stankovic, H. Attar, L. Stankovic, and S. Cheng, "Relay-assisted Rateless Layered Multiple Description Video Delivery," *IEEE Journal on Selected Areas in Communications*, August 2013.